Communications in Computer and Information Science 2535

Rationale
The CCIS series is devoted to the publication of proceedings of computer science conferences. Its aim is to efficiently disseminate original research results in informatics in printed and electronic form. While the focus is on publication of peer-reviewed full papers presenting mature work, inclusion of reviewed short papers reporting on work in progress is welcome, too. Besides globally relevant meetings with internationally representative program committees guaranteeing a strict peer-reviewing and paper selection process, conferences run by societies or of high regional or national relevance are also considered for publication.

Topics
The topical scope of CCIS spans the entire spectrum of informatics ranging from foundational topics in the theory of computing to information and communications science and technology and a broad variety of interdisciplinary application fields.

Information for Volume Editors and Authors
Publication in CCIS is free of charge. No royalties are paid, however, we offer registered conference participants temporary free access to the online version of the conference proceedings on SpringerLink (http://link.springer.com) by means of an http referrer from the conference website and/or a number of complimentary printed copies, as specified in the official acceptance email of the event.

CCIS proceedings can be published in time for distribution at conferences or as post-proceedings, and delivered in the form of printed books and/or electronically as USBs and/or e-content licenses for accessing proceedings at SpringerLink. Furthermore, CCIS proceedings are included in the CCIS electronic book series hosted in the SpringerLink digital library at http://link.springer.com/bookseries/7899. Conferences publishing in CCIS are allowed to use Online Conference Service (OCS) for managing the whole proceedings lifecycle (from submission and reviewing to preparing for publication) free of charge.

Publication process
The language of publication is exclusively English. Authors publishing in CCIS have to sign the Springer CCIS copyright transfer form, however, they are free to use their material published in CCIS for substantially changed, more elaborate subsequent publications elsewhere. For the preparation of the camera-ready papers/files, authors have to strictly adhere to the Springer CCIS Authors' Instructions and are strongly encouraged to use the CCIS LaTeX style files or templates.

Abstracting/Indexing
CCIS is abstracted/indexed in DBLP, Google Scholar, EI-Compendex, Mathematical Reviews, SCImago, Scopus. CCIS volumes are also submitted for the inclusion in ISI Proceedings.

How to start
To start the evaluation of your proposal for inclusion in the CCIS series, please send an e-mail to ccis@springer.com

Tamer Aldwairi · Hisham Al-Mubaid ·
Oliver Eulenstein
Editors

Bioinformatics and Computational Biology

17th International Conference, BICOB 2025
San Francisco, CA, USA, March 17, 2025
Proceedings

Springer

Editors
Tamer Aldwairi
Temple University
Philadelphia, PA, USA

Hisham Al-Mubaid
University of Houston-Clear Lake
Houston, TX, USA

Oliver Eulenstein
Iowa State University
Ames, IA, USA

ISSN 1865-0929 ISSN 1865-0937 (electronic)
Communications in Computer and Information Science
ISBN 978-3-031-94038-5 ISBN 978-3-031-94039-2 (eBook)
https://doi.org/10.1007/978-3-031-94039-2

This Springer imprint is published by the registered company Springer Nature Switzerland AG
The registered company address is: Gewerbestrasse 11, 6330 Cham, Switzerland

Preface

The Organizing Committee of the 17th International Conference on Bioinformatics and Computational Biology (BICOB 2025) would like to present our sincere gratitude and extend a warm welcome to all authors, presenters, attendees, and organizers of BICOB 2025, sponsored by the International Society for Computers and Their Applications (ISCA).

The BICOB conference has enjoyed excellent participation and attendance for more than 16 years so far. It has been establishing itself and growing as one of the leading events for bioinformatics researchers and practitioners. BICOB offers a unique platform for scientists and researchers to present and discuss their research results, methods, and findings with peers who share similar interests. This year, the BICOB conference was held in San Francisco, USA, on March 17–18, 2025.

BICOB 2025 featured keynote talks by well-known scientists and highly competitive contributed talks. The conference covered a wide range of bioinformatics and computational biology topics, including bioinformatics algorithms, genomics, machine learning applications in bioinformatics, and medical informatics.BICOB 2025 brought together researchers, scientists, practitioners, and attendees from multiple countries worldwide. Participants came from various research institutions, including universities, corporations, and government agencies.

Through a thorough review process, BICOB 2025 maintained high standards and quality. This year, each paper submitted to BICOB 2025 underwent single-blind peer review by three peer review by three international program committee members. The papers were evaluated based on their originality, significance, technical content, application content, and presentation style. BICOB 2025 received 38 papers and through a comprehensive and rigorous review process, 18 papers were accepted, giving an acceptance rate of 47%. Also, BICOB 2025 featured a Best Paper Award which was presented during the conference. The online conference management system, *EasyChair*, was used to automate the submission, reviewing, and refereeing processes. The BICOB 2025 Proceedings are published by Springer. All the BICOB 2025 papers will be available online via the *Springer Nature* site, making them easily retrievable through Google and Google Scholar, and they will also be indexed in Scopus.

The professional work of the international program committee and the sub-reviewers is greatly appreciated. We also want to thank the ISCA president, the board of directors, and the ISCA executive director for their continuing support. Furthermore, we extend our gratitude to all the presenters and attendees for their valuable contributions to the success of BICOB 2025

We enjoyed engaging in presentations and discussions at the conference. All participants were invited to connect and build new friendships within the BICOB family. We sincerely hope that every participant will benefit from the conference.

March 2025

Hisham Al-Mubaid
Tamer Aldwairi
Oliver Eulenstein

Organization

General Chair

Hisham Al-Mubaid	University of Houston – Clear Lake, USA

Program Committee Chairs

Tamer Aldwairi	Temple University, USA
Oliver Eulenstein	Iowa State University, USA

International Program Committee

Tavis Anderson	National Animal Disease Center, USDA-ARS, USA
Abdullah Arslan	Texas A & M University-Commerce, USA
Kun-Mao Chao	National Taiwan University, Taiwan
Scott Emrich	University of Tennessee, USA
Krishnendu Ghosh	College of Charleston, USA
Pawel Gorecki	University of Warsaw, Poland
Matthew Hayes	Xavier University of Louisiana, USA
Danny Krizanc	Wesleyan University, USA
Ion Mandoiu	University of Connecticut, USA
Tin Nguyen	University of Nevada, Reno, USA
Yuri Pirola	Università degli Studi di Milano-Bicocca, Italy
Sing-Hoi Sze	Texas A&M University, USA
Bartek Wilczynski	University of Warsaw, Poland
Jianxin Wang	Central South University, China
Jin Zhang	Washington University in St. Louis
Filip Jagodzinski	Western Washington University
Ka-Chun Wong	City University of Hong Kong, China
Pouyan Nahed	University of Nevada, Las Vegas, USA
Letu Qingge	North Carolina Agricultural and Technical State University, USA
Evaldo Costa	Dell Technologies, Brazil

Contents

Leveraging a Swarm-Based Deep Neural Network for Modeling the JAK/STAT Pathway, Predicting Gene–Gene Interactions, and Identifying Therapeutic Targets in AML

Chinyere Ajonu[1](✉), Robert Grundy[2], Graham Ball[3], Dimitrios Zafeiris[2], David Boocock[1], and Golnaz Shahtahmassebi[1]

[1] Nottingham Trent University, Nottingham NG11 8NS, UK
chinyere.ajonu2019@my.ntu.ac.uk
[2] Intelligent Omics Ltd, Nottingham NG1 1GF, UK
[3] Anglia Ruskin University, Cambridge CB1 1PT, UK

Abstract. Neural computational models, driven by brain-inspired cognitive systems, have transformed the modeling of complex biological processes, offering advanced tools such as artificial and deep neural networks to probe intricate signaling pathways in disease contexts. These methods enable the simulation and in-depth analysis of dynamic interactions within critical signaling pathways, facilitating the identification of key disease drivers as therapeutic targets. Hence, this study investigated the JAK/STAT pathway in acute myeloid leukemia (AML), utilizing a Swarm-based deep neural network with transcriptomic data to predict gene– gene interactions and identify critical drivers as potential therapeutic targets. High-throughput analysis of transcriptomic data, integrated with advanced techniques such as net-work inference filtration, interaction matrix visualization, differential driver analysis, and pathway adherence, revealed key age-specific and pan-age disease drivers via the JAK/STAT signaling pathway as potential therapeutic targets in AML. This approach can reliably identify gene–gene interactions and novel disease drivers as potential therapeutic targets, further, investigate established ones, and guide future research in diseases while highlighting the transformative role of Swarm-based deep neural computation in advancing personalized therapeutic strategies and offering a pathway to more effective, tailored interventions in oncology.

Keywords: Neural Network · Swarm-based Deep Neural Network · Acute Myelocytic Leukemia · Therapeutic Target Discovery

1 Introduction

Neural network models are designed to emulate the function of human neurons, enabling efficient processing and analysis of large, complex datasets. These models adapt by modifying the mathematical connections between their processing units, which form the architecture of the network. This process allows for the classification of cases through

T. Aldwairi et al. (Eds.): BICOB 2025, CCIS 2535, pp. 1–11, 2025.
https://doi.org/10.1007/978-3-031-94039-2_1

the assignment of numerical weight values to each input, dynamically adjusting these weights as the model processes data to learn the best solution [1]. Swarm intelligence, a bionetwork inspired computational method that mimics the behavior of deep neural networks (DNNs), has gained popularity and is increasingly integrated with computational models to enhance learning efficiency and adaptability. The incorporation of swarm intelligence optimizes search processes across large datasets, offering advantages such as fault tolerance, scalability, and robust generalization capabilities. These features make artificial neural networks (ANNs), particularly those enhanced with swarm intelligence, highly effective for analyzing unlearned data and tackling complex problems [2–4]. In systems biology, this capability is further expanded through the modeling of signaling pathways, which regulate crucial processes such as cell growth and survival. By applying neural networks to these pathways, researchers can simulate interactions within these networks, uncovering gene-gene interactions and critical molecular drivers as potential therapeutic targets, thus facilitating the development of targeted therapeutic interventions for diseases such as cancer [5].

Acute myeloid leukemia (AML) is a highly aggressive hematologic malignancy characterized by the rapid proliferation of abnormal myeloid precursor cells, which disrupt normal blood cell development. Its heterogeneity, driven by diverse genetic mutations and epigenetic alterations, poses significant challenges to therapeutic intervention and patient outcomes [6]. Central to AML pathophysiology is the Janus kinase/signal transducer and activator of transcription (JAK/STAT) pathway, which plays a key role in regulating critical cellular processes such as proliferation, survival, differentiation, and apoptosis. Dysregulation of this pathway in AML, often through mutation or aberrant activation, leads to uncontrolled cell growth and resistance to apoptosis, contributing to disease progression. Given its importance, the JAK/STAT pathway presents promising therapeutic opportunities, with molecular drivers such as JAK1 emerging as potential targets for personalized treatments [7]. Nevertheless, despite progress in AML therapies, high relapse rates remain a concern, underscoring the need for innovative therapeutic approaches [8]. Hence, using transcriptomic data and a high-throughput swarm-based deep neural network approach, this study aimed to investigate the JAK/STAT pathway's involvement in AML, identify gene-gene interactions, and identify molecular drivers as therapeutic targets.

2 Method

2.1 Data Source, Extraction and Preparation

Transcriptomic profiles from both adult and juvenile AML patients were collected via publicly available datasets from the National Cancer Institute Data Portal and The National Centre for Biotechnology Information. Juvenile AML data were sourced from the TARGET [9] and GSE35784 [10] datasets, whereas adult AML data were sourced from The Cancer Genome Atlas Program (TCGA) [11] and BEAT AML 1.0 [12] datasets. Strict criteria were applied to select untreated primary AML tumors from both the younger and adult cohorts. All patients had a confirmed AML diagnosis, and samples were collected with informed consent, per the Declaration of Helsinki. The inclusion criteria specified bone marrow and peripheral blood blast counts exceeding 50%, with juvenile cohorts ranging from 1 to 29 years of age, and adult cohorts ranging from 21 to 88 years of age.

These criteria resulted in varying cohort sizes across the datasets. The TARGET AML dataset included 187 samples with an average age of 9, and the GSE35784 dataset comprised 130 samples with an average age of under 10. In comparison, the adult TCGA dataset included 151 samples with an average age of 55 years, and the BEAT AML 1.0 dataset contained 350 samples with an average age of 53 years. This broad distribution of data across both age groups enabled comprehensive cross-sectional analyses. The names, symbols, and aliases of genes involved in the JAK/STAT pathway were identified and cataloged via GeneCards. For the GSE35784 dataset, the Affymetrix probe sets corresponding to these genes were determined through GeneAnnot. The 155 elements of the JAK/STAT pathway, as defined by the KEGG and GSEA databases, were extracted along with their corresponding expression values and normalized for each dataset to a range of [0, 1] via min-max normalization, resulting in four distinct datasets used as predictors in the analysis.

2.2 The Backbone SDNN Architecture

Swarm intelligence mimics the behavior of DNNs by utilizing iterative processes that parallel how DNNs leverage multiple hidden layers to refine learning. Instead of relying on increasing the number of hidden layers, as in DNNs, swarm intelligence enhances the search for optimal solutions through repeated, decentralized interactions among smaller, individual models [13]. A swarm-based deep neural network (SDNN) algorithm was employed to model the JAK/STAT signaling pathway, where each known pathway element was sequentially used as a predictor to identify the genes most closely associated with these components. The architecture of the SDNN consists of a three-layer feed-forward perceptron model, structured with the input layer containing the total number of genes (N), a hidden layer with two units, and a single output layer (N-2-1 architecture) (see Fig. 1). A back-propagation algorithm was implemented to optimize the model, facilitating the identification of the most influential genes within the system [14].

A Monte Carlo cross-validation (MCCV) approach was employed to divide the data into training, validation, and testing sets, following a 60:20:20 split for 50 bootstraps to provide the most consistent models. This methodology was applied across various randomized datasets, subsets, and metadata inquiries, ensuring robustness and precision. Throughout the process, datasets were randomly partitioned into subsets, with the collaborative swarm optimizing and making iterative decisions. By repeatedly applying this process across multiple randomized dataset splits, the model enhances its performance and adaptability, utilizing swarm intelligence to optimize search processes. The key learning parameters included 3000 epochs with a termination factor of 1000 epochs, allowing for early stopping when no further improvement was detected. The mean square error (MSE) threshold was set at 0.01, with an MSE window of 100 epochs to ensure accuracy. A learning rate of 0.1, momentum of 0.5, and Pearson's correlation coefficient (r) cutoff of 0.7 were employed to filter relevant features. For each variable, the process was repeated 10 times using 10 distinct swarms as illustrated in Fig. 1, ensuring reliability and consistency in the model's output [14].

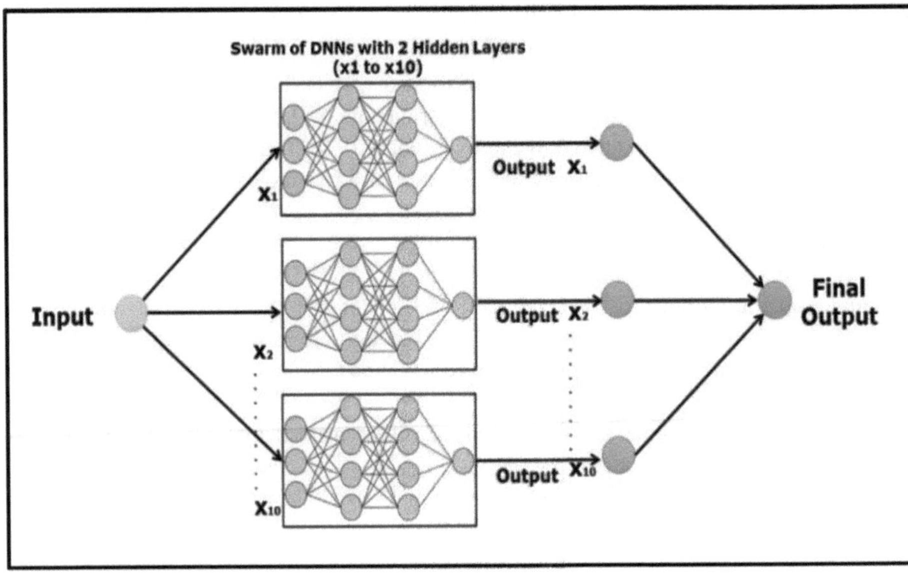

Fig. 1. Schematic illustration of the swarm-based deep neural network used in this study. Figure 1 is taken from Ajonu, et al. 2025 unchanged and licensed under: https://creativecommons.org/licenses/by/4.0/.

For each dataset, the resulting score matrix, generated from the ten randomizations, was averaged to calculate the total strength of each gene-gene interaction. For age-specific analyses, interactions from the adult cohorts (TCGA and BEAT1.0) were subsequently merged and averaged, whereas the GSE35784 and TARGET cohorts were similarly combined. Using the absolute values of the adult and juvenile AML scores, the top 500 gene interactions with the highest absolute values were selected for visualization via Cytoscape version 3.10.2. This visualization approach facilitated the creation of an interactive network map (interactome), where nodes represent sources and targets, and edges illustrate the relationships between them. The thickness of the edges indicates the interaction strength, whereas different colors highlight the interaction type—red for inhibitory (negative) and blue for stimulatory (positive). The arrows show the directionality of the pathway from the source to the target (see Figs. 1 and 2).

2.3 Differential Driver and Pathway Adherence Analysis

Differential driver analysis was conducted to quantify and rank the influence of individual genes within the JAK/STAT pathway on each age group and across all groups based on their overall network impact. This process involves summing the weights that each source gene confers on its targets and vice versa, allowing genes to be ranked based on their influence. Drivers were defined as genes exerting a significant impact on others, and the SDNN framework facilitated iterative quantification of each gene's effect on gene expression levels, enabling the development of low-bias criteria based on cumulative network impact. This approach emphasized genes that may have smaller but consistent effects, offering a more comprehensive view of potential drivers of AML that might otherwise be overlooked via traditional methods. Differential driver analysis was performed across different age-specific cohorts, with median influence values calculated for each gene in both the pediatric and adult groups and across all groups to obtain the pan-age drivers. The genes were subsequently ranked according to their cumulative median influence scores, with those showing the highest network impact in their respective groups being prioritized. Comparative analysis across these cohorts helped uncover age-specific and pan-age drivers of AML, enhancing our understanding of disease mechanisms across all and different age groups.

For the pathway adherence analysis, the stability of the identified drivers was further evaluated by averaging R values derived from 50 bootstrap iterations across the 10 interactions, ensuring robust estimations of model variability and generalizability. This bootstrap method involves generating multiple subsets of the data through random sampling with replacement, which allows for a reliable estimation of how well the model fits the data. Higher R values indicate good fit and stronger adherence, suggesting that a gene closely conforms to its expected role within the pathway. This analysis was crucial in identifying drivers with strong pathway adherence, offering valuable insights into their therapeutic potential and effectiveness. By highlighting those genes with greater adherence, pathway adherence analysis helps prioritize targets for therapeutic intervention, improving the understanding of their role in disease progression and informing the development of more precise treatments.

3 Results

3.1 JAK/STAT Pathway Interaction Models

The iterative process continued until every gene in the datasets underwent quantification yielding a total of 23,870 ($n(n-1)$) associations with 10 repeats. Overall, the distribution of the average of the ten interaction values revealed a predominant occurrence of negative/downregulating interactions across all cohorts, with interaction scores among the four datasets ranging from -12.3 to 2.6. Interactions from the adult cohorts were merged and averaged, whereas those from the younger population were similarly combined for visualization. The top 500 interactions were then selected by filtering based on the largest absolute values and visualized in Figs. 2 and 3 via Cytoscape. In interactomes, the thickness of the edge denotes the strength or weight of the interaction, with red edges indicating negative or downregulating interactions. The direction of the arrow indicates the flow of the interaction from the source (influencing) to the target (influenced) gene. The size of each node served as a representation of the amount of influence each gene exerted on the model as a whole.

The models consistently highlighted unique and differentially influential genes, each of which plays a unique role in the regulatory dynamics of the studied biological systems as evidenced by their large nodes specifically, within the adult cohort, genes such as CREBBP, JAK1, PIAS2, PIK3CA, and IFNL1 emerged as notable drivers (Fig. 2). On the other hand, the younger cohort presented a different set of influential genes, with CBLB, CCND3, IFNA8, EP300, and SPRED2 standing out as the top 5 key drivers in shaping the molecular interactions within this patient group (Fig. 3). Notably, CBLB and CREBBP emerged as overlapping influential genes between the two patient groups. This differential influential gene profile underscores potential age-specific and pan-age molecular mechanisms that may contribute to the observed distinctions in disease characteristics between AML populations. It further presents likely critical interactions influencing disease progression and through the network model, it becomes possible to identify how they should be stimulated, or knocked down, for maximal impact.

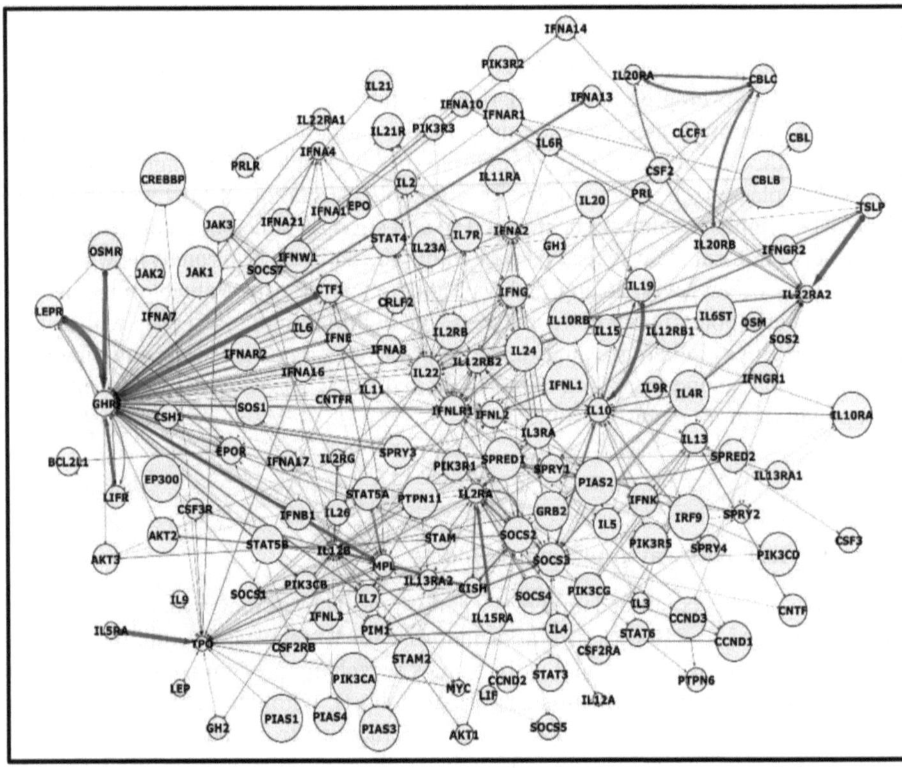

Fig. 2. The organic layer interactome model for the top 500 gene-gene interactions in the JAK/STAT pathway specifically tailored to the adult AML datasets from TCGA and BEAT1.0.

3.2 Differential Driver Analysis and Pathway Adherence

Through a meticulous evaluation of each gene's impact on all other genes and the individual extent of influence, differential driver analysis provides a complementary comprehensive understanding of the network dynamics and underlying mechanisms, tailored specifically to each AML cohort. The analysis outcomes presented in Table 1 are a list of the top 20 genes, according to the ranking of the amount of influence each gene has on the JAK/STAT pathway across all utilized patient groups. In alignment with the interactome presented in Figs. 2 and 3, CBLB, EP300, and CREBBP emerged as the three strongest

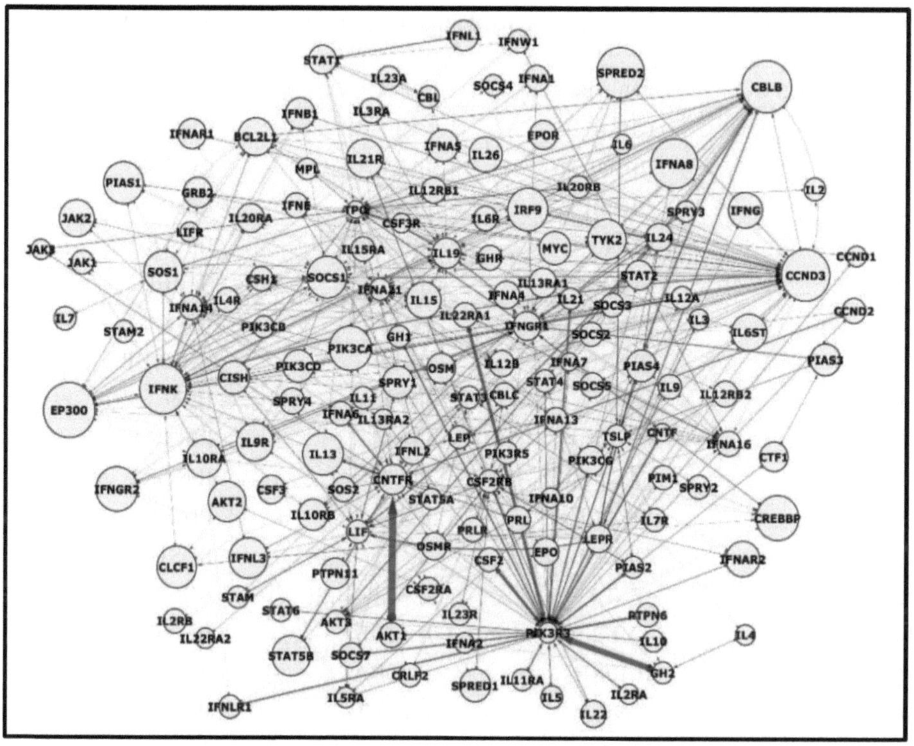

Fig. 3. The organic layer interactome model for the top 500 gene-gene interactions in the JAK/STAT pathway specifically tailored to the AML datasets from TARGET and GSE35784.

drivers of AML, in the JAK/STAT pathway across all patient groups, as shown in Table 1. For the differential examination of adult and younger AML patients, the investigation revealed distinct driver profiles. Within the adult cohort, CREBBP, JAK1, and PIAS2 emerged as the primary drivers of the JAK/STAT pathway (see Table 1). In contrast, the younger cohort exhibited distinct key drivers, with CBLB, CCND3, and EP300 playing prominent roles in steering the AML within the same pathway (see Table 1). These genes significantly influence pathway interactions, underscoring their pivotal roles in driving the biological processes associated with AML in patient groups.

To assess the adherence of the top drivers in the JAK/STAT pathway in AML, we calculated the R values derived from 50 bootstraps across each of the ten iterations of the interaction algorithm, discarding any R values below 0.5 to ensure reliability by eliminating weaker correlations. Note that R values close to 0.7 and above are considered good adherence because they indicate a strong and statistically significant correlation, indicating confidence in the predictive relevance of the data. The top drivers tended to exhibit greater adherence in younger cases, with CCND3 having the highest average R value of 0.80, followed by CBLB, with an R value of 0.78, and IFNA8, with the lowest R value of 0.60 (refer to Fig. 4). Conversely, a notable trend emerged in adult cases, where most top drivers appeared significantly dysregulated, consistently deviating from

Table 1. The top 20 pan-Age, adult, and paediatric AML drivers via the JAK/STAT pathway.

Gene Ranking	Pan-Age	Adult	Paediatric
1	CBLB	CREBBP	CBLB
2	EP300	JAK1	CCND3
3	CREBBP	PIAS2	EP300
4	CCND3	PIK3CA	IFNA8
5	PIK3CA	IFNL1	SPRED2
6	IRF9	EP300	IFNK
7	PIAS1	CBLB	CREBBP
8	IL10RA	IRF9	IRF9
9	IL6ST	PIAS1	PIK3CA
10	SOS1	PIAS3	IFNGR2
11	PIK3CD	IL4R	PIAS1
12	STAT5B	TYK2	SOS1
13	IFNL1	IL6ST	SOCS1
14	TYK2	IL10RB	IFNL3
15	IL10RB	IL10RA	STAT5B
16	IFNK	PIK3CD	IL13
17	SPRED2	IL24	JAK2
18	IL21R	CCND3	IL21R
19	IFNAR1	PTPN11	AKT2
20	JAK2	IFNAR1	IL10RA

pathway adherence, with R values less than 0.70. Despite this, PIAS2 demonstrated relative adherence, being the highest and closest to 0.70, with an R value of 0.69.

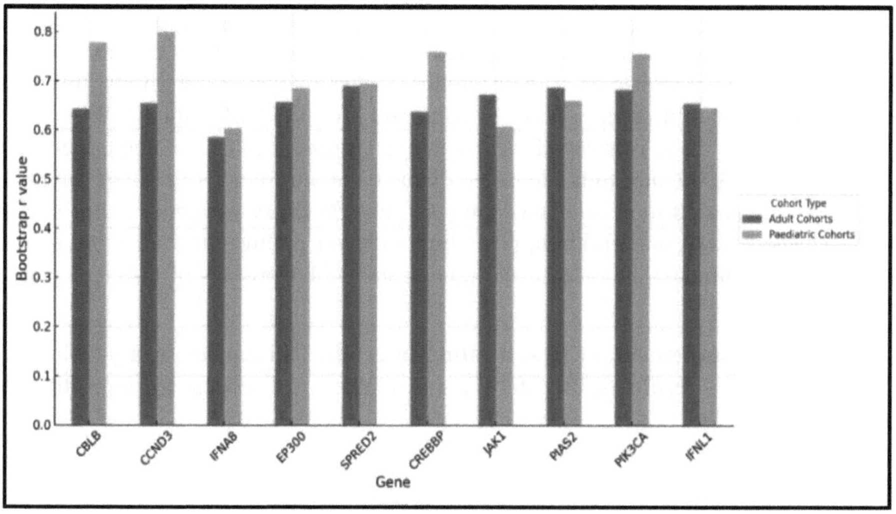

Fig. 4. A clustered column chart illustrating the average bootstrap R-values associated with each top driver of the two patient groups.

4 Conclusion

In this study, an SDNN approach [14] was employed to investigate the JAK/STAT pathway in AML to identify gene-gene interactions and disease drivers as potential therapeutic targets. By leveraging swarm intelligence, the SDNN model enhances the search for optimal solutions, offering an efficient and scalable approach for analyzing large transcriptomic datasets. This method successfully identified important therapeutic targets, such as JAK1 [7], CREBBP [15], EP300 [16], CCND3 [17], and CBLB [18] which are known to be associated with AML pathogenesis and therapy resistance. Importantly, inhibitors targeting JAK1 [19] and CREBBP/EP300 [20] are already in development, highlighting the relevance of these findings in ongoing therapeutic strategies and further demonstrating the robustness and superiority of the methodologies applied. Additionally, the identification of novel AML drivers such as PIAS2, SPRED2, PIK3CA, IFNA8, and IFNL1, highlights new potential therapeutic targets. Both the adult and younger cohort interactomes predominantly showed negative interactions as reflected by the red edges, which aligns with previous research indicating that the JAK/STAT pathway regulates the downstream signaling of several membrane proteins, including G-protein-coupled receptors, and integrins [21]. This approach can consistently identify new differential disease drivers while further investigating established ones, and it can serve as a guide for future research in AML. The findings underscore the effectiveness of using SDNN in complex biological data analysis, offering improved precision in identifying gene-gene interactions and disease-specific drivers hence, contributing to the advancement of therapeutic target discovery in diseases while guiding effective treatment strategies in oncology.

This work has also contributed to the increasing understanding of the molecular differences between adult and pediatric AML [9], reinforcing the importance of personalized therapeutic strategies. The differential driver analysis in this study highlights important molecular differences between adult and juvenile AML patients. While there is some overlap in the key drivers across both age groups, such as pan-age targets such as CBLB and EP3000, the study identified distinct age-specific drivers suggesting that, while there are shared molecular mechanisms, the regulatory dynamics differ significantly between adults and children, providing a clearer pathway toward more precise and effective treatments, tailored to the unique molecular profiles of different patient subgroups.

Availability of Supplementary Results and Materials. All supplementary tables containing the detailed results of this study are available upon request by contacting the corresponding author.

Disclosure of Interests. The authors have no competing interests to declare that are relevant to the content of this article with respect to the research, authorship, or publication of this article.

References

1. Kufel, J., et al.: What is machine learning, artificial neural networks and deep learning?—Examples of practical applications in medicine **13**, 2582 (2023). https://doi.org/10.3390/diagnostics13152582
2. Fang, Y., Wang, Z., Gomez, J., Datta, S., Khan, A.I., Raychowdhury, A.: A swarm optimization solver based on ferroelectric spiking neural networks **13**, 855 (2019). https://doi.org/10.3389/fnins.2019.00855
3. Niu, W., Feng, Z., Feng, B., Xu, Y., Min, Y.: Parallel computing and swarm intelligence based artificial intelligence model for multi-step-ahead hydrological time series prediction **66**, 102686 (2021). https://doi.org/10.1016/j.scs.2020.102686
4. Emambocus, B.A.S., Jasser, M.B., Amphawan, A.: A survey on the optimization of artificial neural networks using swarm intelligence algorithms. Access **11**, 1280–1294 (2023). https://doi.org/10.1109/ACCESS.2022.3233596
5. Prasasya, R.D., Tian, D., Kreeger, P.K.: Analysis of cancer signaling networks by systems biology to develop therapies **21**, 200–206 (2011). https://doi.org/10.1016/j.semcancer.2011.04.001
6. Marcucci, G., Haferlach, T., Döhner, H.: Molecular genetics of adult acute myeloid leukemia: prognostic and therapeutic implications **29**, 475–486 (2011). https://doi.org/10.1200/JCO.2010.30.2554
7. Venugopal, S., Bar-Natan, M., Mascarenhas, J.O.: JAKs to STATs: a tantalizing therapeutic target in acute myeloid leukemia **40**, 100634 (2020). https://doi.org/10.1016/j.blre.2019.100634
8. Bataller, A., et al.: Poor outcome of patients with acute myeloid leukemia in first relapse underscores the need for effective initial treatment **142**, 1480 (2023). https://doi.org/10.1182/blood-2023-179245
9. Bolouri, H., et al.: The molecular landscape of pediatric acute myeloid leukemia reveals recurrent structural alterations and age-specific mutational interactions **24**, 103–112 (2018). https://doi.org/10.1038/nm.4439

10. Jo, A., Mitani, S., Ichikawa, H.: Geo accession viewer, national center for biotechnology information (2019). https://www.ncbi.nlm.nih.gov/geo/query/acc.cgi?acc=GSE3578. Accessed 19 Dec 2023
11. Ley, T.J., et al.: Genomic and epigenomic landscapes of adult de novo acute myeloid leukemia **368**, 2059–2074 (2013). https://doi.org/10.1056/NEJMoa1301689
12. Tyner, J.W., et al.: Functional genomic landscape of acute myeloid leukaemia **562**, 526–531 (2018). https://doi.org/10.1038/s41586-018-0623-z
13. Darwish, A., Hassanien, A.E., Das, S.: A survey of swarm and evolutionary computing approaches for deep learning. Artif. Intell. Rev. **53**, 1767–1812 (2020). https://doi.org/10.1007/s10462-019-09719-2
14. Ajonu, C.I., Grundy, R.I., Ball, G.R., Zafeiris, D.: Application of a high-throughput swarm-based deep neural network algorithm reveals SPAG5 downregulation as a potential therapeutic target in adult AML. Funct. Integr. Genomics **25**, 8 (2025). https://doi.org/10.1007/s10142-024-01514-9
15. LeBlanc, F., Bennett, J., Choi, K., Starczynowski, D.T.: Targeting CREB-binding protein (CREBBP) overcomes resistance to azacitidine and venetoclax therapy in acute myeloid leukemia (AML) **142**, 5765 (2023). https://doi.org/10.1182/blood-2023-187063
16. Wu, W., et al.: The epigenetic regulators EP300/CREBBP represent promising therapeutic targets in MLL-rearranged acute myeloid leukemia **10**, 206 (2024). https://doi.org/10.1038/s41420-024-01940-5
17. Smith, C.C., et al.: Recurrent mutations in cyclin D3 confer clinical resistance to FLT3 inhibitors in acute myeloid leukemia **27**, 4003–4011 (2021). https://doi.org/10.1158/1078-0432.CCR-20-3458
18. Caligiuri, M.A., et al.: Novel c-CBL and CBL-b ubiquitin ligase mutations in human acute myeloid leukemia **110**, 1022–1024 (2007). https://doi.org/10.1182/blood-2006-12-061176
19. Pemmaraju, N., et al.: A phase I/II study of the janus kinase (JAK)1 and 2 inhibitor ruxolitinib in patients with relapsed or refractory acute myeloid leukemia **15**, 171–176 (2015). https://doi.org/10.1016/j.clml.2014.08.003
20. Knurowski, T., et al.: An open-label phase I/IIa study to evaluate the safety and efficacy of CCS1477 as monotherapy and in combination in patients with advanced hematological malignancies **83**, CT136 (2023). https://doi.org/10.1158/1538-7445.AM2023-CT136
21. Villarino, A.V., Kanno, Y., Ferdinand, J.R., O'Shea, J.J.: Mechanisms of jak/STAT signaling in immunity and disease **194**, 21–27 (2015). https://doi.org/10.4049/jimmunol.1401867

Modeling Glucose-Insulin Dynamics with Delay Differential Equations

Fathalla A. Rihan[1(✉)], K. Udhayakumar[1], M-Naim Anwar[2], and Mutaz Mohammad[3]

[1] Department of Mathematical Sciences, College of Science, UAE University, Al Ain 15551, UAE
{frihan,udhayakumar_k}@uaeu.ac.ae
[2] Department of Mathematics, Faculty of Engineering, Pharos University in Alexandria, Alexandria, Egypt
mohamed.naeem@pua.edu.eg
[3] Department of Mathematics, Zayed University, Abu Dhabi 144534, UAE
Mutaz.Mohammad@zu.ac.ae

Abstract. This study proposes a delay differential model for glucose-insulin endocrine and metabolic regulation. The model involves two time delays, δ_g and δ_ι, which represent delayed insulin secretion and delayed glucose reduction. Moderate hyperglycemia results in beta-cell growth (negative feedback), whereas severe hyperglycemia results in beta-cell reduction (positive feedback). Hopf bifurcation occurs when a time delay passes bifurcation points. Based on biological findings, the model exhibits periodic oscillations. Diabetes and pre-diabetes may be characterized by chaotic glucose-insulin dynamics, which makes blood sugar levels unpredictable and difficult to control. The theoretical results have been validated by numerical simulations.

Keywords: Delay differential equations · Glucose-insulin · Stability · Chaos behavior · Hopf bifurcation

1 Introduction

Insulin is a hormone that transports glucose from the bloodstream to cells where it can be used for energy production [9]. In response to blood glucose levels, $\beta-$cells in the pancreas release insulin [18]. Due to imbalances in insulin and glucose interactions, diabetes mellitus poses a significant challenge to public health worldwide. Many scientists have investigated the mechanisms involving insulin, glucose, and $\beta-$cells [15,20]. Over the past few decades, several mathematical models of glucose-insulin regulating systems have been developed and analyzed [3,21]. In terms of studying glucose-insulin dynamics, [3] has been regarded as a pioneer. In order to mimic blood glucose regulation, he developed a simple model. A variety of mathematical models based on Bolie's findings have been used to study glucose-insulin regulation. [6] proposes a dynamical model' with delay. The authors presented positive and bounded solutions to the stability of unique equilibrium. Due to the delay, the dynamical model overcomes the limitations of the coupled minimal model. You can find more related studies at [8,11,17,19] and their references.

T. Aldwairi et al. (Eds.): BICOB 2025, CCIS 2535, pp. 12–25, 2025.
https://doi.org/10.1007/978-3-031-94039-2_2

Incorporating time delay into mathematical models is a novel strategy for addressing glucose-insulin dynamics. Lags, or simply delays, indicate the time it takes for complex biological processes to occur [7]. Using time-delay models in biological modeling is becoming more common [7, 16]. Several of these models have been used in chemostat theory [22], epidemiology [5], circadian rhythms, neural networks, and genetic regulatory networks [10]. In the research work [2, 12], time delays were incorporated into the glucose-insulin feedback system. Chuedoung et al. [4] developed a one-compartment model to analyze the oscillations of the glucose and insulin dynamic system. Two specific time delays were investigated in this study. The first, $\delta_{\mathcal{G}}$, relates to insulin production delayed by glucose levels, and the second, $\delta_{\mathcal{I}}$, pertains to hepatic glucose response delayed by insulin levels. They hypothesized that the combined impact of two-time delays influenced glucose-insulin regulation dynamics rather than each delay individually. To understand the dynamics of glucose-insulin interactions, time-delayed mathematical models were developed.

In several studies, deterministic models have been used to explore glucose-insulin dynamics. Our study presents two types of time delays in glucose-insulin interactions. We show how time delays influence system stability and bifurcation. Section 2 of the manuscript presents the formulation of the deterministic model, followed by its qualitative analysis, stability analysis, and bifurcation analysis. Numerical simulations are presented in Sect. 3. Section 4 summarizes conclusions and future directions

2 Modeling of Glucose-Insulin Interactions

We first investigate the analysis of deterministic model of glucose-insulin system. The-model system consists of three distinct dynamical variables: $\mathcal{I}(t)$ represents the concentration of insulin at time t [mins] that is higher than its basal level, $\mathcal{G}(t)$ represents the concentration of glucose that is higher than its basal level, based on Bajaj [1], $\mathcal{B}(t)$ represents the number of beta-cells. Herein, we present the following delayed deterministic model of glucose-insulin regulation:

$$\begin{aligned}\dot{\mathcal{I}}(t) =& -\varrho_2\mathcal{I}(t) + \varrho_1\mathcal{G}(t-\delta_{\mathcal{G}})\mathcal{B}(t-\delta_{\mathcal{G}}) + \varsigma_1\mathcal{B}(t-\delta_{\mathcal{G}}),\\ \dot{\mathcal{G}}(t) =& \varsigma_2 - \varrho_4\mathcal{I}(t-\delta_{\mathcal{I}}) + \varrho_3 N/\mathcal{B}(t),\\ \dot{\mathcal{B}}(t) =& -\varrho_7\mathcal{B}(t) - \varrho_8\mathcal{G}^2(t)\mathcal{B}(t) + \varrho_6\mathcal{B}(t)(\kappa - \mathcal{B}(t)) + \varrho_5(\mathcal{G}(t) - \hat{\mathcal{G}})(\kappa - \mathcal{B}(t)),\end{aligned} \tag{1}$$

with the initial history conditions of the model system (1) is satisfied within the interval $[-\delta_{\max}, 0]$, where $\delta_{\max} = \max\{\delta_{\mathcal{G}}, \delta_{\mathcal{I}}\}$ as defined in the space

$$\mathscr{C}_+ = \{\psi \in \mathscr{C}([t_0 - \delta_{\max}, t_0], \mathbb{R}^3_+) : \mathcal{I}(\vartheta) = \psi_1(\vartheta), \mathcal{G}(\vartheta) = \psi_2(\vartheta), \mathcal{B}(\vartheta) = \psi_3(\vartheta)\} \tag{2}$$

where $\psi = (\psi_1(\vartheta), \psi_2(\vartheta), \psi_3(\vartheta)) \in C([t_0 - \max\{\delta_{\mathcal{G}}, \delta_{\mathcal{I}}\}, t_0], \mathbb{R}^3_+)$, the Banach space of all interval continuous mappings $[-\delta_{\max}, 0]$ to $\mathbb{R}^3_+$ for $\psi_i(\vartheta) \geq 0 (\vartheta \in [-\delta_{\max}, 0], i = 1, 2, 3, 4)$ and $(\psi_1(0), \psi_2(0), \psi_3(0)) > 0$, with $\mathbb{R}^3_+ = \{(\mathcal{I}, \mathcal{G}, \mathcal{B}) : \mathcal{I} > 0, \mathcal{G} > 0, \mathcal{B} > 0\}$.

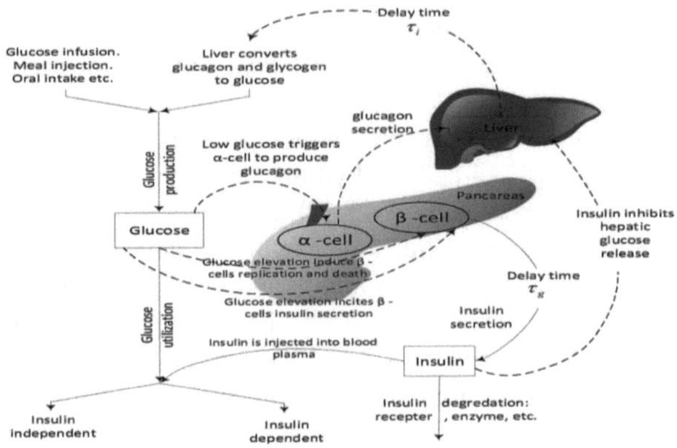

Fig. 1. Glucose and insulin metabolism compartmentalized; see [23].

Insulin, as a hormone, degrades similarly to other hormones. In addition to its function in converting excess glucose into glycogen, it is employed in various other processes that ultimately result in its degradation. The rate of change in the amount of insulin above its basel level is represented in the first equation of (1). Here, the term $\varrho_1\mathcal{G}(t-\delta_\mathcal{G})\mathcal{B}(t-\delta_\mathcal{G})$ refers to the increase in insulin from a plasma glucose spike with a time delay $\delta_\mathcal{G}$; the decreasing rate of insulin-independent of glucose is $-\varrho_2\mathcal{I}(t)$; $\varsigma_1\mathcal{B}(t-\delta_\mathcal{G})$ denotes the increase in plasma insulin concentration that beta-cells produce at a time delay of $\delta_\mathcal{G}$. See Fig. 1. In the second equation, the $\varrho_3 N/\mathcal{B}(t)$ is added because the glucose concentration in the plasma depends upon the N/n capacity of the beta-cells. The rate at which glucose concentration levels decrease in response to insulin release at a time delay $\delta_\mathcal{I}$ is denoted by $\varrho_4\mathcal{I}(t-\delta_\mathcal{I})$. The third equation in (1) represents the rate of change in the number of beta-cells. The parameter κ in the model equation represents the total density of beta-cells, and $\hat{\mathcal{G}}$ represents glucose metabolism in its basal state. As blood glucose levels rise beyond the fasting level and beta cells are not actively dividing, $\varrho_5(\mathcal{G}(t)-\hat{\mathcal{G}})(\kappa-\mathcal{B}(t))$ increases the division of beta-cells. $\varrho_6\mathcal{B}(t)(\kappa-\mathcal{B}(t))$ denotes the rate at which beta-cell number increases due to the bilateral interaction between dividing and non-dividing beta-cells and the reduction in $\mathcal{B}(t)$ caused by the beta-cell current level is represented by the term $\varrho_7\mathcal{B}(t)$. The rate of replication of beta-cell mass increases as the glucose level rises. However, in cases of severe hyperglycemia, beta-cell replication may be decreased. Beta-cell mortality in vitro has been demonstrated to fluctuate non-linearly with glucose concentration. By increasing the blood glucose concentration in the medium surrounding cultured beta-cells from 0 to approximately 11 mM, the mortality rate for those cells decreased. Beta-cell mortality rates remain low or rise when glucose concentration exceeds 11 mM. The expression $\varrho_8\mathcal{G}^2(t)\mathcal{B}(t)$ modifies the growth rate of beta-cells so that it follows a logistic pattern.

There are two significant time delays in the model (1): the time between changes in blood glucose levels and the liver's physiological responses to control them:

(a) The glucose-triggered insulin production delay $\delta_\mathcal{G}$ occurs when blood glucose levels rise, typically after eating. As a result of insulin, glucose is taken up by the liver and stored as glycogen. Responses are not instantaneous, however. As a result, there is a time delay before the liver can effectively respond to changes in glucose levels, which arise from several physiological and biochemical processes. Based on the clinical data presented by Palumbo et al. [13], it was found that the time delay, $\delta_{\mathcal{G}}$, in insulin production in a healthy individual in response to increased glucose levels were measured to be 23.5 min.

(b) Hepatic glucose response delay $\delta_{\mathcal{I}}$. As blood glucose levels drop, glucagon leaves the pancreas, which prompts the liver to release glucose again. To maintain glucose homeostasis, the liver processes hormonal signals and executes metabolic adjustments in response to changes in blood glucose levels over the hepatic glucose response delay $\delta_{\mathcal{I}}$. As insulin levels fluctuate, Prager et al. [14] report significant variation in the hepatic glucose level adjustment time delay $\delta_{\mathcal{I}}$.

As a result of incorporating two-time delays into the glucose-insulin regulatory system, it is more realistic and represents glucose-insulin biological regulation over time more accurately. When time delays are included in the differential model, the stability solution and model dynamics are enhanced.

Lemma 2.1 *(Existence and Uniqueness). If the initial values of the model (1) can be chosen as* $(\mathcal{I}(0) > 0, \mathcal{G}(0) > 0, \mathcal{B}(0) > 0)$, *then there exists a unique solution in* $[0, \infty)$ *and positive for all* $t \geq 0$.

Proof: The model system of (1) can be rewritten in the form:

$$\dot{\mathcal{X}} = \mathcal{F}(\mathcal{X}) \tag{3}$$

where,

$$\dot{\mathcal{X}} = (\dot{\mathcal{I}}(t), \dot{\mathcal{G}}(t), \dot{\mathcal{B}}(t))^T \in \mathbb{R}^3,$$

$$\mathcal{F}(\mathcal{X}) = \begin{pmatrix} -\varrho_2\mathcal{I}(t) + \varrho_1\mathcal{G}(t-\delta_{\mathcal{G}})\mathcal{B}(t-\delta_{\mathcal{G}}) + \varsigma_1\mathcal{B}(t-\delta_{\mathcal{G}}) \\ \varsigma_2 - \varrho_4\mathcal{I}(t-\delta_{\mathcal{I}}) + \varrho_3 N/\mathcal{B}(t) \\ -\varrho_7\mathcal{B}(t) - \varrho_8\mathcal{G}^2(t)\mathcal{B}(t) + \varrho_6\mathcal{B}(t)(\kappa - \mathcal{B}(t)) + \varrho_5(\mathcal{G}(t) - \hat{\mathcal{G}})(\kappa - \mathcal{B}(t)) \end{pmatrix}$$

where the mapping $\mathcal{F} : \mathscr{C}_+ \to \mathbb{R}^3$ and $\mathcal{F} \in C^\infty(\mathbb{R}^3)$. Let $\mathcal{F}(\vartheta) = (\psi_1(\vartheta), \psi_2(\vartheta), \psi_3(\vartheta)) \in \mathscr{C}_+$ and $\psi_i(\vartheta) \geq 0$, for all $i = 1, 2, 3, 4$ with $\vartheta \in [-\delta_{\max}, 0]$, where $\delta_{\max} = \max\{\delta_{\mathcal{G}}, \delta_{\mathcal{I}}\}$. It is evident that the vector function $\mathcal{F}(\mathcal{X})$ is a continuous and locally Lipschitz function of the variables $\mathcal{I}, \mathcal{G}, \mathcal{B}$ in the domain $\Omega = \{(\mathcal{I}(t), \mathcal{G}(t), \mathcal{B}(t)) : \mathcal{I} > 0, \mathcal{G} > 0, \mathcal{B} > 0\}$. Thus, any solution $(\mathcal{I}, \mathcal{G}, \mathcal{B})$ of the model system (1) with initial history condition (2) is exists and unique in the interval $[-\delta_{\max}, 0]$, for all $t \geq 0$.

Lemma 2.2 *(Positiveness). For any* $t > 0$, *all of the solutions in the model (1) exist and are positive.*

Proof: Assume that the solution of (1) is $(\mathcal{I}(t), \mathcal{G}(t), \mathcal{B}(t))$. For all $t_0 > 0$, if $\mathcal{B}(t_0) = 0$, then $\mathcal{B}'(t_0) \leq 0$.Since $\mathcal{B}(t_0) = 0$, we have $\mathcal{B}'(t_0) = -\varrho_7\mathcal{B}(t_0) - \varrho_8\mathcal{G}^2(t)\mathcal{B}(t_0) + \varrho_6\mathcal{B}(t_0)(\kappa - \mathcal{B}(t_0)) + \varrho_5(\mathcal{G}(t_0) - \hat{\mathcal{G}})(\kappa - \mathcal{B}(t_0)) = \varrho_5(\mathcal{G}(t_0) - \hat{\mathcal{G}})\kappa > 0$ at t_0. This contradiction shows $\mathcal{B}(t) > 0$ for all t within the interval of existence. If $\mathcal{I}(\bar{t}_0) = 0$ for some $\bar{t}_0 > 0$, then $\mathcal{I}'(\bar{t}_0) \leq 0$ and $0 \geq \mathcal{I}'(\bar{t}_0) = -\varrho_2\mathcal{I}(\tilde{t}_0) + \varrho_1\mathcal{G}(\tilde{t}_0 - \delta_{\mathcal{G}})\mathcal{B}(\tilde{t}_0 - \delta_{\mathcal{G}}) + \varsigma_1\mathcal{B}(\tilde{t}_0 - \delta_{\mathcal{G}}) = \varrho_1\mathcal{G}(\tilde{t}_0 - \delta_{\mathcal{G}})\mathcal{B}(\tilde{t}_0 - \delta_{\mathcal{G}}) + \varsigma_1\mathcal{B}(\tilde{t}_0 - \delta_{\mathcal{G}})$. Since $\mathcal{B}(\tilde{t}_0 - \delta_{\mathcal{G}}) > 0$ and thus $\mathcal{I}(t) > 0$ for all t in the interval of existence.

2.1 Equilibrium Points

The equilibrium points of (1) are respectively $\mathcal{I}^*, \mathcal{G}^*, \mathcal{B}^*$. Hence by putting $\dot{\mathcal{I}} = 0, \dot{\mathcal{G}} = 0, \dot{\mathcal{B}} = 0$ in (1), one can obtain

$$\begin{aligned} 0 =& -\varrho_2\mathcal{I}^* + \varrho_1\mathcal{G}^*\mathcal{B}^* + \varsigma_1\mathcal{B}^*, \\ 0 =& \varsigma_2 - \varrho_4\mathcal{I}^* + \varrho_3 N/\mathcal{B}^*, \\ 0 =& -\varrho_7\mathcal{B}^* - \varrho_8\mathcal{G}^{2*}\mathcal{B}^* + \varrho_6\mathcal{B}^*(\kappa - \mathcal{B}^*) + \varrho_5(\mathcal{G}^* - \hat{\mathcal{G}})(\kappa - \mathcal{B}^*), \end{aligned} \tag{4}$$

From second equation of (4),

$$\mathcal{B}^* = \frac{\varrho_3 N}{\varsigma_2 - \varrho_4\mathcal{I}^*} \tag{5}$$

Substituting it into first equation of (4)

$$\mathcal{I}^* = \frac{\mathcal{A}}{2} \pm \frac{\sqrt{\mathcal{A}^2 - 4\mathcal{B}}}{2}, \quad \mathcal{A} = \frac{\varsigma_2\varrho_2}{\varrho_2\varrho_4}, \mathcal{B} = \frac{\varrho_3 N(\varrho_4\mathcal{G}^* + \varsigma_1)}{\varrho_2\varrho_4} \tag{6}$$

where $\mathcal{G}^*$ is the positive root of the given equation

$$\begin{aligned} &\left(\mathcal{A}^2 - 4\mathcal{B} + 4\left(\mathcal{A}_1^2 - 4\mathcal{B}_1\right)\varsigma_2^2\right)^2 + \left(4\varrho_3 N - \mathcal{A}\left(1 + 2\varsigma_2\right)\right)^4 \\ &-2\left(\mathcal{A}^2 - 4\mathcal{B} + 4\left(\mathcal{A}_1^2 - 4\mathcal{B}_1\right)\varsigma_2^2\right)\left(4\varrho_3 N - \mathcal{A}\left(t + 2\varsigma_2\right)\right)^2 = 16\left(\mathcal{A}^2 - 4\mathcal{B}\right)\left(\mathcal{A}_1^2 - -4\mathcal{B}_1\right)\varsigma_2^2 \end{aligned} \tag{7}$$

with

$$\mathcal{A}_1 = \frac{1}{\varrho_6}\left(\varrho_7 + \varrho_8\mathcal{G}^* - \varrho_6\kappa + \varrho_5\mathcal{G}^* + \varrho_5\hat{\mathcal{G}}\right), \mathcal{B}_1 = \frac{\varrho_5\kappa(\hat{\mathcal{G}} - \mathcal{G}^*)}{\varrho_6}.$$

2.2 Stability and Bifurcation Analysis

The aim of this section is to determine whether the delayed system (1) is stable at any given point $E\left(\mathcal{I}_e, \mathcal{G}_e, \mathcal{B}_e\right)$. The delayed system (1) has been linearized at point E, which is given by

$$\frac{d\Lambda(t)}{dt} = \Delta_1\Lambda(t) + \Delta_2\Lambda(t - \delta_{\mathcal{G}}) + \Delta_3\Lambda(t - \delta_{\mathcal{I}}) \tag{8}$$

where,

$$\Delta_1 = \begin{pmatrix} -\varrho_2 & 0 & 0 \\ 0 & 0 & \varrho_3(\frac{N}{\mathcal{B}_e^2}) \\ 0 & M_1 & M_2 \end{pmatrix}, \Delta_2 = \begin{pmatrix} 0 & \varrho_1\mathcal{B}_e & (\varrho_1\mathcal{G}_e + \varsigma_1) \\ 0 & 0 & 0 \\ 0 & 0 & 0 \end{pmatrix}, \Delta_3 = \begin{pmatrix} 0 & 0 & 0 \\ -\varrho_4 & 0 & 0 \\ 0 & 0 & 0 \end{pmatrix},$$

$$M_1 = \varrho_5(\kappa - \mathcal{B}_e) - \varrho_8(2\mathcal{G}_e\mathcal{B}_e), M_2 = \varrho_r(\kappa - 2\mathcal{B}_e) + \varrho_5(\hat{g} - \mathcal{G}_e) - \varrho_7 - \varrho_8\mathcal{G}_e^2$$

and

$$\Lambda(\cdot) = (\mathcal{I}(\cdot), \mathcal{G}(\cdot), \mathcal{B}(\cdot))^T$$

Then the appropriate variational matrix of the linearized system (8) at any point $E\left(\mathcal{I}_e, \mathcal{G}_e, \mathcal{B}_e\right)$ is expressed as

$$\mathcal{J}_E = \begin{pmatrix} w_{11} & w_{12}e^{-\omega\delta_{\mathcal{G}}} & w_{13}e^{-\omega\delta_{\mathcal{G}}} \\ w_{21}e^{-\omega\delta_{\mathcal{I}}} & 0 & w_{23} \\ 0 & w_{32} & w_{33} \end{pmatrix} \tag{9}$$

where

$$\begin{aligned} w_{11} =& -\varrho_2; w_{21} = \varrho_1\mathcal{B}_e; w_{31} = \varrho_1\mathcal{G}_e + \varsigma_1; \\ w_{21} =& -\varrho_4; w_{23} = \varrho_3(\frac{N}{\mathcal{B}_e^2}); \quad w_{32} = M_1; w_{33} = M_2. \end{aligned}$$

Subsequently, the characteristic equation is formulated as

$$\mathrm{Det}(\Delta_1 + \Delta_2 e^{-\omega\delta_{\mathcal{G}}} + \Delta_3 e^{-\omega\delta_{\mathcal{I}}} - \omega\mathbb{I}_3) = 0, \tag{10}$$

where $\mathbb{I}_3$ denotes the third-order identity matrix. The characteristic equation above results in

$$\Theta(\omega, \delta_{\mathcal{G}}, \delta_{\mathcal{I}}) = P_1(\omega) + P_2(\omega)e^{-\omega\delta_{\mathcal{I}}-\omega\delta_{\mathcal{G}}} \tag{11}$$

where

$$P_1(\omega) =\omega^3 + l_1\omega^2 + l_2\omega + l_3, \quad P_2(\omega) = l_4\omega + l_5.$$

with

$$\begin{aligned} l_1 =& -M_2 + \varrho_2; l_2 = -M_2\varrho_2 - \frac{NM_1\varrho_3}{\mathcal{B}^2}; l_3 = -\frac{NM_1\varrho_2\varrho_3}{\mathcal{B}^2}; \\ l_4 =& \mathcal{B}\varrho_1\varrho_4; l_5 = \mathcal{G}M_1\varrho_1\varrho_4 - \mathcal{B}M_2\varrho_1\varrho_4 + M_1\varrho_4\varsigma_1. \end{aligned}$$

We note that Eq. (11) is transcendental in ω. We will examine the local stability analysis of both the delayed and non-delayed systems at the equilibrium point E by analyzing the sign of the real parts of the roots of (11). The given equilibrium E is locally asymptotically stable if all the roots of (11) are negative or possess negative real parts. The transcendental equation that governs our system will have infinite solutions. Thus, the Routh-Hurwitz criteria fail to provide a local stability analysis for the delayed system. Therefore, we analyze the local stability of the non-delayed system using Routh-Hurwitz criteria as follows:

Case 1: $\delta_{\mathcal{G}} = \delta_{\mathcal{I}} = 0$ If there is no delay in the hepatic glucose response and the glucose-triggered insulin production, we can rewrite the characteristics equation (11) as follows:

$$\omega^3 + l_1\omega^2 + (l_2 + l_4)\omega + (l_3 + l_5) = 0. \tag{12}$$

The traditional Routh-Hurwitz criteria will be utilized to ascertain the sign of the roots of the characteristic equation (12). Thus, we have following results At the equlibrium state E, the model system (1) is said to be locally asymptotically stable, if the following conditions are hold:

$$l_1 > 0, l_2 + l_4 > 0, l_3 + l_5 > 0 \text{ and } l_1(l_2 + l_4) > l_3 + l_5 \tag{13}$$

Case 2: $\delta_{\mathcal{G}} > 0, \delta_{\mathcal{I}} = 0$. Let us consider the glucose-triggered insulin production delay and there is no delay in the hepatic glucose response, then we can rewrite the characteristics equation (11) as follows:

$$\Theta(\omega, \delta_{\mathcal{G}}, 0) = P_1(\omega) + P_2(\omega)e^{-\omega\delta_{\mathcal{G}}} = 0 \tag{14}$$

The characteristic equation (14) reduced to the equation (12) when the time delay $\delta_{\mathcal{G}} = 0$ and the Routh-Hurwitz criteria (13) holds, in this case, the equlibrium point E is locally asymptotically stable. However, our objective is to examine the system (14) in the context of time delay, specifically when $\delta_{\mathcal{G}} > 0$. If the roots of equation (14) cross the imaginary axis, the stability around the equilibrium point E will vary. Now putting $\omega = i\alpha(\alpha > 0)$ into the system (14) we separate its real and imaginary parts, we have the following equations:

$$\begin{aligned} &-\alpha^2 l_1 + l_3 = l_5 \cos\alpha\delta_{\mathcal{G}} - \sin\alpha\delta_{\mathcal{G}}\alpha l_4; \\ &-\alpha^3 + \alpha l_2 = \alpha l_4 \cos\alpha\delta_{\mathcal{G}} - \sin\alpha\delta_{\mathcal{G}} l_5. \end{aligned} \tag{15}$$

After squaring and combining the above two equations, we get

$$\alpha^6 + \alpha^4\theta_1 + \alpha^2\theta_2 + \theta_3 = 0. \tag{16}$$

where

$$\theta_1 = -l_1^2 - 2l_2, \theta_2 = l_2^2 + 2l_1 l_3 + l_4^2, \theta_3 = l_5^2 - l_3^2$$

When $\mu = \alpha^2$, (16) becomes

$$\mathscr{F}(\mu) = \mu^3 + \mu^2\theta_1 + \mu\theta_2 + \theta_3 = 0. \tag{17}$$

If suppose $l_5^2 - l_3^2 < 0$, then $\mathscr{F}(0) < 0$ and $\mathscr{F}(+\infty) = +\infty$, this indicates that there is at least one positive root in Eq. (17). This means that there is a μ_0 is the only positive root that satisfies equation (17). Thus, Eq. (16) possesses at least one positive root, referred to as α_0. Since α_0 is a root of the system (16), hence the characteristic equation (14) has a pair of completely imaginary roots of the type $\pm i\alpha_0$. By solving both equations of (15), we obtain

$$\tan(\alpha\delta_{\mathcal{G}}) = \frac{(\alpha^2 l_1 + l_3)(\alpha l_4) + (\alpha l_2 - \alpha^3)l_5}{(\alpha^2 l_1 + l_3)l_5 + (\alpha l_2 - \alpha^3)(\alpha l_4)} \tag{18}$$

Then, $\delta_{\mathcal{G}}^a$ corresponding to α_0 and for $a = 0, 1, 2, \ldots$ is represented by

$$\delta_{\mathcal{G}}^a = \frac{1}{\alpha_0}\arctan\left(\frac{(\alpha^2 l_1 + l_3)(\alpha l_4) + (\alpha l_2 - \alpha^3)l_5}{(\alpha^2 l_1 + l_3)l_5 + (\alpha l_2 - \alpha^3)(\alpha l_4)}\right) + \frac{2\pi a}{\alpha_0} \tag{19}$$

For the bifurcation parameter $\delta_{\mathcal{G}}^{a} = 0$, the equilibrium point E will exhibit local asymptotic stability if the condition (13) are satisfied. Moreover, according to, we could say that E will exhibit stability for $\delta_{\mathcal{G}} < \delta_{\mathcal{G}}^{a}$.

To investigate the Hopf bifurcation, we will check the transversality condition.

$$\left[\frac{d(\operatorname{Re}\omega)}{d\delta_{\mathcal{G}}}\right]_{\delta_{\mathcal{G}}=\delta_{\mathcal{G}}^{a}} > 0. \tag{20}$$

It suggests the presence of at least one eigenvalue with a positive real component, satisfying $\delta_{\mathcal{G}} > \delta_{\mathcal{G}}^{a}$. It is necessary to check the transversality condition.

We give the following assumption:

$$(\mathbf{A1}) \qquad 2\alpha_0^4 l_4 > 2l_1 l_5 \alpha_0^2$$

Lemma 2.3. *If* $\omega(\delta_{\mathcal{G}}) = \mu(\delta_{\mathcal{G}}) + i\alpha(\delta_{\mathcal{G}})$ *is the root of (14) at* $\delta_{\mathcal{G}} = \delta_{\mathcal{G}}^{a}$ *which satisfies* $\mu(\delta_{\mathcal{G}}^{a}) = 0, \alpha(\delta_{\mathcal{G}}^{a}) = \alpha_0$, *then* $\operatorname{Re}\left(\frac{d\omega}{d\delta_{\mathcal{G}}}\right)\Big|_{\delta_{\mathcal{G}}=\delta_{\mathcal{G}}^{a},\alpha=\alpha_0} > 0$

Proof: From (14)

$$\left[(3\omega^2 + 2l_1\omega + l_2) + l_4 e^{-\omega\delta_{\mathcal{G}}} - (l_4\omega + l_5)\delta_{\mathcal{G}} e^{-\omega\delta_{\mathcal{G}}}\right]\frac{d\omega}{d\delta_{\mathcal{G}}} = \omega(l_4\omega + l_5)e^{-\omega\delta_{\mathcal{G}}} \tag{21}$$

which implies that

$$\left(\frac{d\omega}{d\delta_{\mathcal{G}}}\right)^{-1} = \frac{3\omega^2 + 2l_1\omega + l_2}{-\omega(\omega^3 + l_1\omega^2 + l_2\omega + l_3)} + \frac{l_4}{\omega(l_4\omega + l_5)} - \frac{\delta_{\mathcal{G}}}{\omega} \tag{22}$$

Then

$$\begin{aligned}
\operatorname{Re}\left[\left(\frac{d\omega}{d\delta_{\mathcal{G}}}\right)^{-1}\right] &= \operatorname{Re}\left[\frac{3\omega^2+2l_1\omega+l_2}{-\omega(\omega^3+l_1\omega^2+l_2\omega+l_3)} + \frac{l_4}{\omega(l_4\omega+l_5)} - \frac{\delta_{\mathcal{G}}}{\omega}\right]_{\omega=i\alpha_0}, \\
&= \operatorname{Re}\left[\frac{(-3\alpha_0^2+l_2)+i(2l_1\alpha_0)}{(l_2\alpha_0^2-\alpha_0^4)+i(l_1\alpha_0^3-l_3\alpha_0)} + \frac{l_4}{(-l_4\alpha_0^2)+i(l_5\alpha_0)} - \frac{\delta_{\mathcal{G}}}{i\alpha_0}\right], \\
&= \frac{2\alpha_0^4 l_4 - 2l_1 l_5\alpha_0^2}{(l_2\alpha_0^2-\alpha_0^4)(-l_4\alpha_0^2)-(l_1\alpha_0^3-l_3\alpha_0)(l_5\alpha_0)}
\end{aligned} \tag{23}$$

By **(A1)**, one has

$$\operatorname{Re}\left[\left(\frac{d\omega}{d\delta_{\mathcal{G}}}\right)^{-1}\right]\Bigg|_{\delta_{\mathcal{G}}=\delta_{\mathcal{G}}^{a},\alpha=\alpha_0} > 0. \tag{24}$$

This completes the proof.

Theorem 2.4. *For the model system (1), if* $\delta_{\mathcal{I}} = 0$, *and* $\delta_{\mathcal{G}} > 0$ *and Routh-Hurwitz criteria (13), (A1) hold, then the given equilibrium point is locally asymptotically stable when* $\delta_{\mathcal{G}} \in [0,\ \delta_{\mathcal{G}}^{a})$ *and a Hopf bifurcation occurs around the equilibrium when* $\delta_{\mathcal{G}} = \delta_{\mathcal{G}}^{a}$.

3 Numerical Results and Discussions

In this section, we discuss the dynamical analysis of glucose-insulin interactions numerically in the presence of two time delays. To begin our analysis, we will first explore the deterministic aspects of the glucose-insulin interactions within the proposed model. According to Table 1, numerical simulations have been performed using a range of parameter values. Simulations are performed using continuous RK method [16] for (1) with step size $h = 0.01$. Herein, we present numerical simulations to study the impact of time delays on glucose-insulin dynamics using a mathematical model (1). It is important to mention that time delay influences the equilibrium stability. Although it is possible to get chaotic dynamics with different parameter choices, in this paper, our objective is to analyze how varying delay parameters influence the system's behavior. To investigate the influence of the time delays on the dynamics of the system (1), it is necessary to determine the corresponding bifurcation point. The values of bifurcation points can be calculated using equations (19). When considering the time delays of insulin secretion stimulated by glucose level and glucose reduction with respect to the increase of insulin concentration is zero ($\delta_{\mathcal{G}} = \delta_{\mathcal{I}} = 0$), we determine that $l_1 = 0.9090 > 0; l_3 + l_5 = 0.3575 > 0; l_1(l_2 + l_4) - l_3 + l_5 = 0.2813 > 0$, implying that according to the Routh-Hurwitz criterion the system (1) is asymptotically stable with respect to the steady state E.

Table 1. The parameters used in the glucose-insulin interaction model (1)

Parameter	Meaning	Value
ϱ_1	increasing rate of insulin concentration	0.48 min^{-1}
ϱ_2	decreasing rate of insulin concentration	0.25 min^{-1}
ϱ_3	decreasing rate in insulin secretion due to	
	a reduction to $\frac{1}{N}$ of the normal number n of the beta-cell cells	0.83 min^{-1}
ϱ_4	reduction rate of glucose concentration	0.58 min^{-1}
ϱ_5	increasing rate of dividing beta-cell cells	0.45 min^{-1}
ϱ_6	rate of increase of beta-cell cells number	0.31 min^{-1}
ϱ_7	rate of reduction of beta-cell cells number	0.36 min^{-1}
ϱ_8	rate of beta-cell cells mass replication	0.012 min^{-1}
ς_1	increasing rate of plasma insulin level	0.1 min^{-1}
$\hat{g}$	difference between glucose base level and glucose fasting level	1.4 mg/dl
κ	total density of beta-cell cells number	1.5 mg/dl
$\delta_{\mathcal{G}}$	delay in insulin secretion caused by glucose	0–5 min
$\delta_{\mathcal{I}}$	delay in glucose reduction caused by an increase in insulin	0–5 min

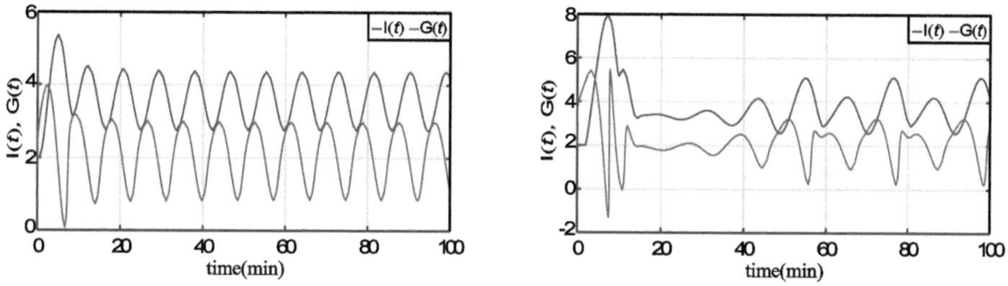

(a) $\delta_G = 1.5, \delta_I = 0.5$

(b) $\delta_G = 2.5, \delta_I = 1$

Fig. 2. The state trajectories of glucose-insulin in (1) with time delays.

The Impact of Time Delays $\delta_{\mathcal{G}}$ and $\delta_{\mathcal{I}}$ on Glucose-Insulin Dynamics: When we consider that the time delays of insulin secretion stimulated by glucose level and glucose reduction with respect to the increase in insulin concentration is positive, that is, $\delta_{\mathcal{G}}, \delta_{\mathcal{I}} > 0$ we simulated the time series and phase plots of the insulin and glucose dynamics in Figs. 2 and 3. Figures 2(a) and 3(b) show the state trajectories of the glucose-insulin interactions with different sets of time delays ($\delta_{\mathcal{G}} = 1.5, \delta_{\mathcal{I}} = 0.5$) and $\delta_{\mathcal{G}} = 2.5, \delta_{\mathcal{I}} = 1$, respectively. From Fig. 2(a), we can see that the glucose (red line) and insulin (blue line) concentrations exhibit periodic oscillations at a relatively high frequency. This figure shows a quick response of insulin to changes in glucose levels with a shorter delay $\delta_{\mathcal{G}} = 1.5$. Moreover, the concentrations of both glucose and insulin oscillations appeared to be moderate, indicating a relatively stable interaction between glucose and insulin levels with mild variations over time. When the time delays were considerably increased, the frequency of oscillations in both glucose and insulin concentrations decreased significantly, as shown in Fig. 2(b). This means that the system reacts more slowly to changes in glucose levels. In addition, the height of the glucose oscillations increased, while the amplitude of the insulin oscillations appeared somewhat more irregular. This indicated a less stable interaction with glucose levels. From the simulations of Figs. 2, we can see that a short but reasonable delay i.e., $\delta_{\mathcal{G}} = 1.5$ in insulin release allows the body to avoid overcompensation for minor fluctuations in blood glucose levels. That is, the delay provides the system time to evaluate whether the rise in glucose is a temporary spike or a sustained increase. This ensures that small spikes in glucose do not lead to excessive insulin secretion, which could cause hypoglycemia (low blood sugar). If the delay ($\delta_{\mathcal{G}} = 2.5, \delta_{\mathcal{I}} = 1$) is too long, especially if we increase the delay $\delta_{\mathcal{G}}$, the insulin response becomes slow, which can lead to prolonged hyperglycemia (high blood sugar levels) and decreased effectiveness in controlling blood sugar levels. This delayed response is often observed in individuals with insulin resistance and type 2 diabetes. Figure 3(a) shows the corresponding time evaluation of insulin and glucose concentration above its base level at a time delay $\delta_g = 1.2$ (min) and $\delta_\iota = 0.05$ (min). Figure 3(b) shows the corresponding time evaluation of beta-cell number and glucose concentration above its base level at the time delay $\delta_g = 1.2$ (min) and $\delta_\iota = 0.05$ (min). For the parameters listed in Table 1, a periodic solution is obtained after 10 min. A similar periodic behavior for physiological fluctuations of glucose and insulin was demonstrated. These oscillations represent the

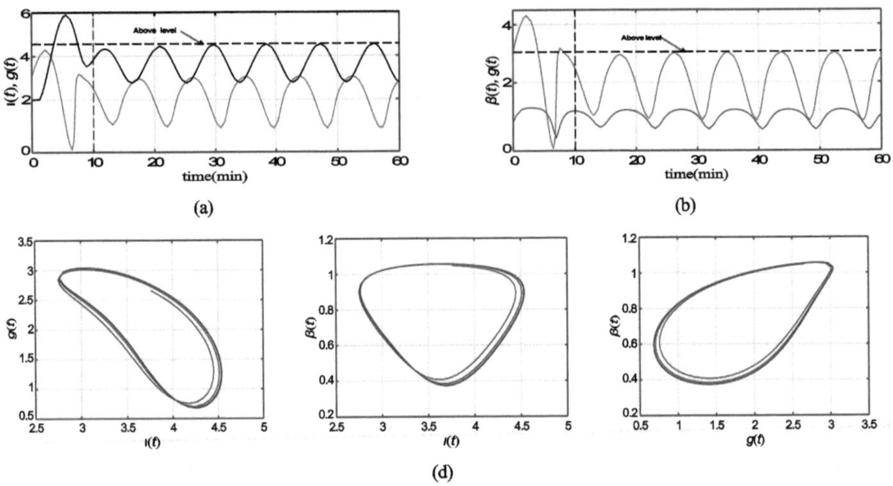

Fig. 3. The state trajectories and phase portrait position of the elements of glucose-insulin regulatory system (1) with time delay $\delta_g = 1.2, \delta_\iota = 0.05$ (min) (a) State trajectories of glucose (mg/dl) (red) and insulin $(\mu U/ml)$ (black) concentration. (b) State trajectories of glucose (mg/dl) (red) and beta-cells number (blue). (d) Phase portrait position of the elements of glucose-insulin regulatory system. (Color figure online)

dynamic nature of insulin-glucose feedback loops, and periodically modulated signals are more effective in target cells than static, random, or chaotic signals.

Bifurcation Diagrams: To investigate different patient scenarios, we adjusted two important values: the hepatic glucose response delay $\delta_{\mathcal{G}} \in [1, 4]$ and glucose-triggered insulin production delay $\delta_{\mathcal{I}} \in [1, 4]$. A one-dimensional bifurcation diagram is a useful tool that is generally used to reveal the attractor type, to which the system dynamics finally settle down after the transient phase, to show the impact of individual time delay on the system behavior, and to provide more information about the dependence of the system dynamics on certain parameters. The simulation of the system (1) demonstrates that both regular and irregular periodic oscillations occur, often known as chaotic behavior, as illustrated in the one-dimensional bifurcation diagram, and are plotted in Figures 4 based on varying the discrete time delays $\delta_{\mathcal{G}}$ and $\delta_{\mathcal{I}}$ as a bifurcation parameter and the other system parameters in Table 1. Figure 4(a) and (c) show how the concentration of insulin, $\mathcal{I}(t)$ and glucose, $\mathcal{G}(t)$, change as the time delay, $\delta_{\mathcal{G}}$ varies, with a relatively low insulin production rate ($\varrho_1 = 0.3$). For $\varrho_1 = 0.3$, the initial region (low values of $\delta_{\mathcal{G}}$) exhibited clear, stable periodic oscillations. These regions showed regular and consistent peaks, indicating that the insulin and glucose levels fluctuated in a stable manner. In both Figs. 4(a) and (c), even amidst chaotic behavior, there are periodic windows where the dynamics temporarily stabilize and exhibit periodic oscillations again. The presence of periodic limit cycles is indicative of natural rhythms in glucose-insulin interactions, maintaining a consistent pattern that helps regulate the body's glucose levels effectively. Figures 4(b) and (d) show the insulin and glucose concentration dynamics with a higher insulin production rate ($\varrho_1 = 0.7$) as the time delay $\delta_{\mathcal{G}}$ varies. Compared with Figs. 4(a)

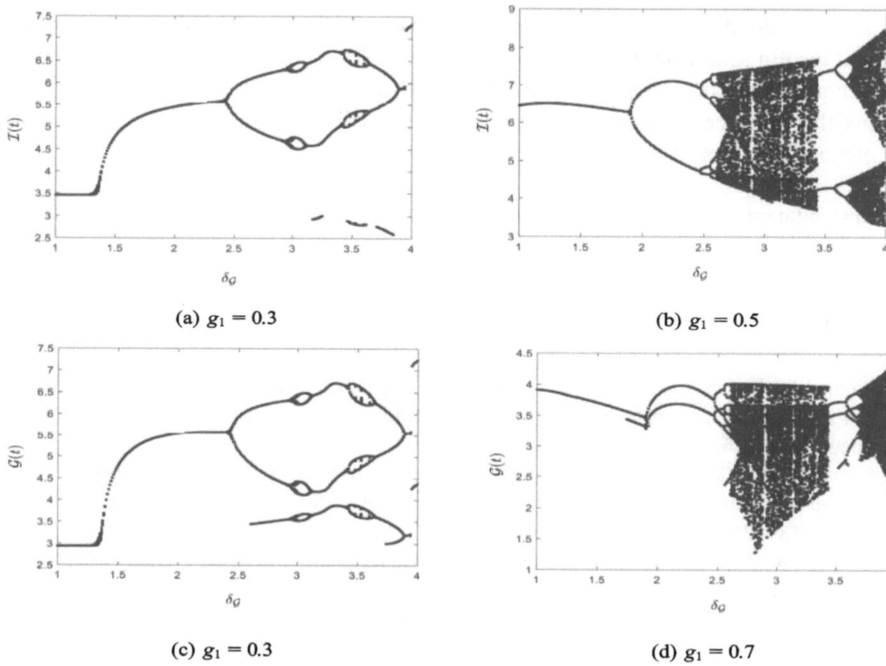

Fig. 4. Bifurcation diagram for the gluscose-insulin dynamics (1) with time delay $\delta_{\mathcal{G}} \in [1, 4]$, $\varrho_1 = 0.3$, $\varrho_1 = 0.7$.

and (c), the system enters a chaotic state at lower values of $\delta_{\mathcal{G}}$, indicating that a higher rate of insulin production destabilizes the response more quickly. That is, for a higher rate of $\varrho_1 = 0.7$, as $\delta_{\mathcal{G}}$ increases, the system enters a chaotic state characterized by dense regions of scattered points. From Figs. 4(b) and (d), we can see the transition from stable periodic oscillations to chaos and the period-doubling cascades occurred. These cascades are seen as successive bifurcations, where the single oscillation splits into two (when $\delta_{\mathcal{G}} = 1.895$), then four (when $\delta_{\mathcal{G}} = 2.463$), and continue to multiply.

4 Conclusions

The glucose-insulin interaction system is complex and requires sophisticated models to depict these interactions accurately. This study examined the effects of time delays on the dynamics of glucose-insulin. The occurrence of chaos in glucose-insulin dynamics can help explain why blood sugar levels in individuals with diabetes or pre-diabetes can be highly unpredictable and difficult to control. Understanding these dynamics may aid in designing more effective treatments that aim to minimize these instabilities. In this paper, the asymptotic stability of the model and its Hopf bifurcation properties are discussed. Asymptotic stability of the delayed fractional-order glucose-insulin model is guaranteed by certain conditions derived from the characteristic equation. The bifurcation parameters are δ_g and δ_ι. Under sufficient conditions, our model also guarantees Hopf bifurcations. Glucose drop delays caused by increased insulin concentration and insulin secretion delays caused by glucose levels can lead to complex dynamics. According to phase portraits and bifurcation diagrams, time delays create chaotic regions that may cause metabolic problems. It will be our future focus to estimate parameters from real observations.

Acknowledgment. The research was funded by the UAE University (UAE), project # 12S141.

References

1. Bajaj, J.S., Rao, G.S., Rao, J.S., Khardori, R.: A mathematical model for insulin kinetics and its application to protein-deficient (malnutrition-related) diabetes mellitus (PDDM). J. Theor. Biol. **126**(4), 491–503 (1987)
2. Bennett, D.L., Gourley, S.A.: Asymptotic properties of a delay differential equation model for the interaction of glucose with plasma and interstitial insulin. Appl. Math. Comput. **151**(1), 189–207 (2004)
3. Bolie, V.W.: Coefficients of normal blood glucose regulation. J. Appl. Physiol. **16**(5), 783–788 (1961)
4. Chuedoung, M., Sarika, W., Lenbury, Y.: Dynamical analysis of a nonlinear model for glucose-insulin system incorporating delays and β-cells compartment. Nonlinear Anal.: Theory Methods Appl. **71**(12), e1048–e1058 (2009)
5. Cooke, K., Van den Driessche, P., Zou, X.: Interaction of maturation delay and nonlinear birth in population and epidemic models. J. Math. Biol. **39**, 332–352 (1999)
6. De Gaetano, A., Arino, O.: Mathematical modelling of the intravenous glucose tolerance test. J. Math. Biol. **40**, 136–168 (2000)
7. Forde, J.E.: Delay differential equation models in mathematical biology. University of Michigan (2005)
8. Gao, C., Chen, F., Yu, P.: Dynamics of a new delayed glucose-insulin model with obesity. Int. J. Bifurcat. Chaos **34**(06), 2450070 (2024)
9. Hedeskov, C.J.: Mechanism of glucose-induced insulin secretion. Physiol. Rev. **60**(2), 442–509 (1980)
10. Lakshmanan, S., Rihan, F.A., Rakkiyappan, R., Park, J.H.: Stability analysis of the differential genetic regulatory networks model with time-varying delays and Markovian jumping parameters. Nonlinear Anal. Hybrid Syst. **14**, 1–15 (2014)
11. Li, C., Liu, Y., Wang, Y., Feng, X.Dynamic modeling of the glucose–insulin system with inhibitors impulsive control. Math. Methods Appl. Sci. (2024)
12. Li, J., Kuang, Y., Mason, C.C.: Modeling the glucose-insulin regulatory system and ultradian insulin secretory oscillations with two explicit time delays. J. Theor. Biol. **242**(3), 722–735 (2006)
13. Palumbo, P., Panunzi, S., De Gaetano, A., et al.: Qualitative behavior of a family of delay-differential models of the glucose-insulin system. Discrete Continuous Dyn. Syst. Ser. B. **7**(2), 399 (2007)
14. Prager, R., Wallace, P., Olefsky, J.M., et al.: In vivo kinetics of insulin action on peripheral glucose disposal and hepatic glucose output in normal and obese subjects. J. Clin. Investig. **78**(2), 472–481 (1986)
15. Quianzon, C.C., Cheikh, I.: History of insulin. J. Commun. Hospit. Intern. Med. Perspect. **2**(2), 18701 (2012)

16. Rihan, F.A.: Delay Differential Equations and Applications to Biology. Springer (2021). https://doi.org/10.1007/978-981-16-0626-7
17. Rihan, F.A., Udhayakumar, K.: Optimal control of glucose-insulin dynamics via delay differential model with fractional-order. Alex. Eng. J. **114**, 243–255 (2025)
18. Rorsman, P., Renstrom, E.: Insulin granule dynamics in pancreatic beta cells. Diabetologia **46**, 1029–1045 (2003)
19. Singh, J., Kamboj, S., Saxena, S.: Analysis of glucose-insulin dynamics as an integrator in type 1 diabetes. In: 2023 International Conference on Sustainable Emerging Innovations in Engineering and Technology (ICSEIET), pp. 663–665. IEEE (2023)
20. Vecchio, I., Tornali, C., Bragazzi, N.L., Martini, M.: The discovery of insulin: an important milestone in the history of medicine. Front. Endocrinol. **9**, 613 (2018)
21. Yang, J., Tang, S., Cheke, R.A.: The regulatory system for diabetes mellitus: modeling rates of glucose infusions and insulin injections. Commun. Nonlinear Sci. Numer. Simul. **37**, 305–325 (2016)
22. Zhao, T.: Global periodic-solutions for a differential delay system modeling a microbial population in the chemostat. J. Math. Anal. Appl. **193**(1), 329–352 (1995)
23. Al-Hussein, A.B.A., Rahma, F., Jafari, S.: Hopf bifurcation and chaos in time-delay model of glucose-insulin regulatory system. Chaos, Solitons Fractals **137**, 109845 (2020). https://doi.org/10.1016/j.chaos.2020.109845. Issn: 0960-0779

Transitive Reduction and Cluster Normalization for Improved Gene Regulatory Network Inference

Youchuan Wang(✉) and Chilukuri K. Mohan

Department of Electrical Engineering and Computer Science, Syracuse University, Syracuse, NY 13244, USA
{ywang552,ckmohan}@syr.edu

Abstract. We propose improvements to algorithms for gene regulation network inference using gene knockout data. In particular, we attempt to reduce the errors caused due to transitive (indirect) connections on the regulation strength. To address this problem, the matrix of connection strengths (between gene pairs) is first modified by thresholding and normalization with respect to other connection strengths, reducing the impact of indirect connections. This is followed by a cluster normalization step that distinguishes low magnitude direct connections from indirect connections. Our approach combines ranking selection, Gaussian process modeling, and thresholding, thereby enhancing the quality of results obtained with gene regulation network inference. When compared with the top-performing Median Corrected Z scores method, we obtained consistent improvements in the well-known Area Under the Precision Recall Curve (AUPRC) metric, across all ten 100-gene networks from the DREAM3 and DREAM4 contest.

Keywords: Transitive Reduction · DREAM Challenge · Median Corrected Z Scores (MCZ) · Gene Knockout Data

1 Introduction

Gene regulatory networks (GRNs) are critical for understanding the complex interactions that govern gene expression within biological systems. They represent how genes regulate one another, influencing various cellular processes such as development, differentiation, and response to environmental stimuli. The inference of GRNs involves reconstructing these networks from high-throughput gene expression data, which is both challenging and vital for advancing our understanding of biological systems and disease mechanisms. Accurately identifying gene-gene interactions provides valuable insights into the biological processes that govern gene expression and regulation.

Modeling techniques in GRN inference enable more precise representation of the underlying biological mechanisms, and facilitate the analysis of complex regulatory relationships. However, accurately inferring these models from experimental data remains a significant challenge due to the inherent noise and high dimensionality of gene expression data.

T. Aldwairi et al. (Eds.): BICOB 2025, CCIS 2535, pp. 26–41, 2025.
https://doi.org/10.1007/978-3-031-94039-2_3

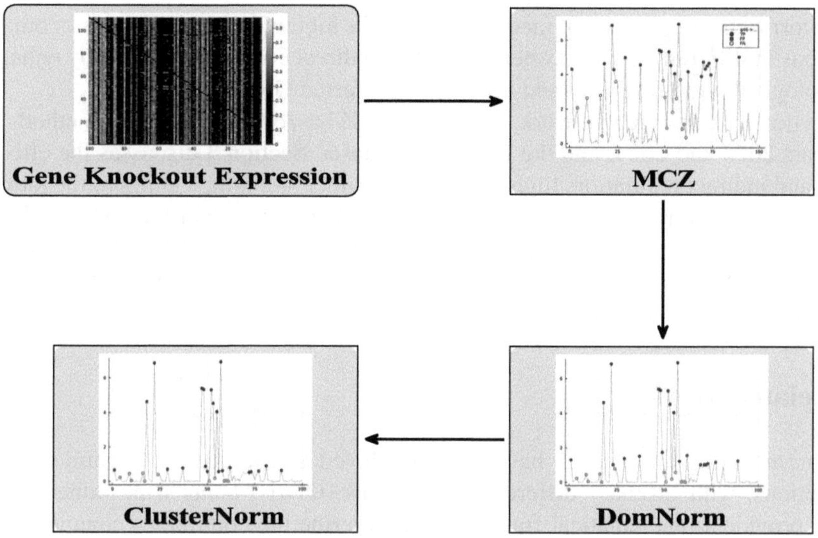

Fig. 1. Four-node pipeline with directed arrows between nodes

The DREAM4 challenge has highlighted the significance of using such models in GRN inference, with participants employing various approaches to reconstruct the regulatory interactions. Among these, the Median Corrected Z-scores (MCZ) method has emerged as a strong performer, achieving high rankings in the competition. Despite its success, the MCZ algorithm has notable limitations, particularly in distinguishing between direct and indirect gene interactions. High correlation values can occur in both direct and indirect connections, leading to a misidentification of regulatory paths and an increase in false positives. For example, gene X may directly influence gene Y, and gene Y may directly influence gene Z, resulting in the indirect influence of X on Z, which may be mistakenly represented as a direct connection from X to Z.

To address the limitations of the original MCZ algorithm, we propose modifications to enhance its precision in identifying biologically meaningful gene regulatory connections. As shown in Fig. 1, our pipeline begins with gene knockout expression data, similar to the original MCZ approach, to calculate regulatory strength between genes. Our first modification, *Dominant Normalization* **(Dom Norm)**, normalizes the weighted adjacency matrix, resulting in a modified matrix that reduces the impact of indirect paths on ranking selection. By attenuating the effect of consecutive paths, Dom Norm provides a more accurate assessment of direct connections. We then apply **Cluster Norm**, a clustering based refinement technique, resulting in a second modified matrix; this step isolates clusters of mid-range connections, distinguishing indirect influences from direct regulatory interactions and improving overall precision in network inference. Our approach also integrates ranking selection, Gaussian process probability, and threshold filtering to increase the robustness of the inferred connections while reducing false positives.

Using ten (five of each) 100-gene networks from the DREAM3 and DREAM4 datasets, we evaluated the performance of our method, showing that it consistently outperforms the original MCZ method. The results highlight the advantages of our modifications in differentiating between direct and indirect connections, thereby enhancing the biological interpretability and accuracy of inferred GRNs.

Section 2 reviews recent work, including MCZ-based GRN inference methods from the years 2023 and 2024, and the DREAM4 dataset. Section 3 discusses the effects of direct and indirect regulation, high and low regulatory strength, leading to the introduction of the Dom Norm and Cluster Norm. Section 4 discusses data and code availability, as well as the performance comparison on the DREAM3 and DREAM4 datasets with five networks of 100 genes from each challenge. Section 5 summarizes the paper and discusses future research directions.

2 Related Work

Various modeling approaches have been employed to capture the dynamics of gene interactions, with ordinary differential equations (ODEs) being a prominent choice. ODEs provide a mathematical framework to describe the continuous changes in gene expression levels over time, taking into account the regulatory influences between genes. A typical ODE model [15] can be expressed as follows:

$$\frac{dX_i}{dt} = f(X) + \sum_{j \in \text{Regulators}} g_{ij}(X_j) \tag{1}$$

In gene regulatory networks (GRNs), let X_i represent the expression level of gene i, where $f(X)$ captures the intrinsic dynamics of gene expression and $g_{ij}(X_j)$ represents the regulatory influence of gene j on gene i. A nonzero $g_{i,j}$ indicates a regulatory link from gene j to gene i within the GRN. The challenge of GRN inference lies in uncovering these regulatory interconnections, specifically determining how the interaction $g_{i,j}$ reflects the actual connections in the network.

In this paper, we use the well-known DREAM3 and DREAM4 [8] gene network datasets, with knockout gene expression (i.e., the gene expression for all genes in the network when one gene is deleted) values. A knockout matrix Y^{KO} of size $n \times n$ is given for a network of m genes. The i'th row of Y^{KO} is the steady state of the network after the knockout of gene i.

$$Y^{KO} = \begin{bmatrix} 0 & ko_{12} & ko_{13} & \cdots \\ ko_{21} & 0 & ko_{23} & \cdots \\ ko_{31} & ko_{32} & 0 & \cdots \\ \vdots & \vdots & \vdots & \ddots \end{bmatrix}$$

We begin with the "wild-type" or unperturbed gene regulation networks. For each network, we have 10 sets of observations corresponding to 10 different unperturbed random initializations (for gene expression values), as well as 1 set of observations that represent equilibrium values. Combining the observations of the wild-type, Y^{WT}, and knockouts, Y^{KO}, we get our full wild-type matrix Y

$$Y = \begin{bmatrix} w_{1,1} & w_{1,2} & w_{1,3} & \cdots \\ w_{2,1} & w_{2,2} & w_{2,3} & \cdots \\ \vdots & \vdots & \vdots & \ddots \\ w_{11,1} & w_{11,2} & w_{11,3} & \cdots \\ 0 & ko_{1,2} & ko_{1,3} & \cdots \\ \vdots & \vdots & \vdots & \ddots \end{bmatrix}$$

In gene regulatory network (GRN) inference, the **gold standard (gs)** refers to an adjacency matrix provided by the DREAM3 and DREAM4 challenge dataset, which contains the true regulatory relationships between genes, serving as a crucial benchmark for evaluating the performance of various inference algorithms. However, this gold standard is not accessible during the inference process; instead, algorithms rely solely on gene expression data, such as wildtype and knockout profiles, to deduce regulatory interactions. By comparing inferred networks against *gs*, researchers can assess the accuracy of their methods and identify false positives and negatives, thus refining their algorithms and enhancing our understanding of gene expression regulation.

The influence of gene interactions in gene regulatory networks can be quantitatively assessed using the following formula [8, 12]:

$$W[i, j] = \left| \frac{ko_{i,j} - \text{median}(\mathbf{Y}_j)}{\text{var}(\mathbf{Y}_j)} \right|$$

In this equation, $W\,[i, j]$ represents the regulatory strength or influence from gene j to gene i. The term $ko_{i,j}$ denotes the knockout data for gene i when gene j is turned off. $\mathbf{Y}_j$ is the column vector of gene expression values for gene j across all experimental conditions. This expression quantifies how much the expression level of gene i deviates from its normal state—represented by the median of its expression levels across a set of conditions—when gene j is knocked out. The difference between the knockout observation $ko_{i,j}$ and median(Y_j), and normalizing by the variance, reflects the extent to which gene i behaves differently in response to the absence of gene j, compared to what would be expected under normal conditions. If the normalized deviation is substantial, it suggests that gene j has a significant regulatory influence on gene i. Therefore, according to the MCZ algorithm, a strong deviation indicates the presence of a regulatory edge from gene j to gene i.

The original MCZ paper proposes four different pipelines for gene regulatory network inference. Among these, MCZ utilizes only wild-type and knockout data, which simplifies the process and enhances speed. In contrast, the other pipeline, termed tlCLR, leverages time-series gene expression data in conjunction with L1-norm regression. The four pipelines represent different combinations of these two methods. Notably, MCZ alone outperforms two pipelines that rely on tlCLR, while the combined approach of tlCLR and MCZ provides only a marginal improvement over MCZ alone. The knockout and wild-type data contain critical information that aids in accurately reconstructing the structure of the gene regulatory network.

The hierarchical clustering techniques employed in [4], which utilize ontological relationships and metrics such as path lengths and least common subsumers (LCS), strongly resonate with our approach using *Cluster Norm* to identify and group regulatory modules in gene networks. Both strategies rely on structuring complex biological data into meaningful clusters, driven by domain-specific relationships. Similarly, the disease clustering work in [3], leveraging functional annotations from the Gene Ontology, mirrors our methodology by grouping genes with shared regulatory functions, showcasing the power of ontology-based approaches in elucidating biological similarities. Moreover, DupTree's gene tree parsimony algorithm [16], designed to optimize reconciliation costs and explore alternate rootings, presents an interesting computational analogy to our study. Both approaches aim to refine the accuracy and efficiency of large-scale biological analyses by focusing on modular relationships and optimizing network characteristics through innovative, relationship-driven methods.

In 2023 and 2024, various MCZ-based methodologies have been proposed for inferring and analyzing gene regulatory networks, each with unique approaches to handle the complexity of biological systems. One approach is a gene ranking and model prioritization framework, which systematically ranks genes based on their regulatory importance, helping researchers to prioritize models that better capture the underlying biological interactions [7]. Another innovative method involves using directed graph convolutional networks (GCNs) to construct networks. This approach leverages deep learning to handle the directed nature of gene interactions, improving the accuracy of inferred connections by capturing both local and global network features [17].

Bayesian inference models gene expression randomness, while fuzzy cognitive maps refine networks by addressing sparsity and non-linear relationships [9, 11]. Factor graph models generalize insights to new experiments with minimal assumptions [5]. Our approach aligns with and complements these advancements. Despite the advancements in alternative methodologies, the "original MCZ method" [8], against which we compare our approach, remains the strongest performer within the DREAM challenge. This highlights its continued relevance and the need for further exploration into methods that can outperform or complement its effectiveness in gene regulatory network inference.

3 Methods

In this section, we address the effects of indirect connections and weakly-expressing direct regulations on the values of MCZ calculations. To address these issues, we propose three algorithms: **Dominant Normalization** and **Cluster Normalization**.

3.1 Direct and Indirect Regulations

We first consider the effects of transitive or indirect influences among genes, with another gene as an intermediary. Figures 2, 3, 4, 5, 6, 7 and 8 present example results on network 1 for the DREAM4 dataset; similar results are obtained for other networks, but they are omitted due to space constraints.

Let the prediction of regulation from gene i to j be denoted by $X_{i,j}$, and the proposed adjacency matrix by X. Denote the i-th column and i-th row of matrix W by $W[:, i]$ and

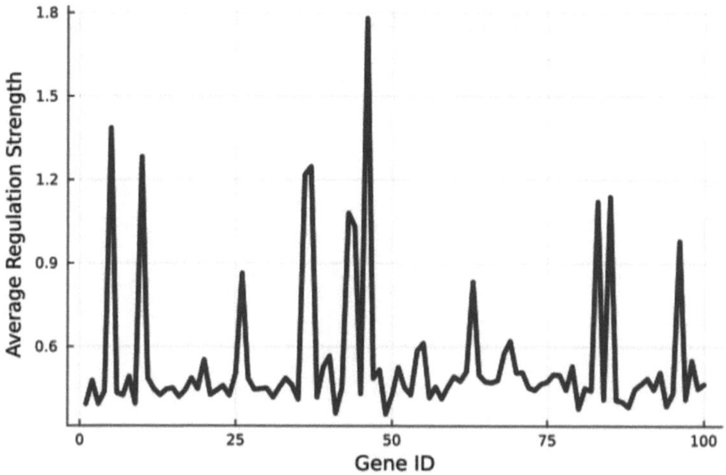

Fig. 2. Average regulatory strength from each gene as the source across all target genes in the network, highlighting the relative influence of each gene on the network's connectivity.

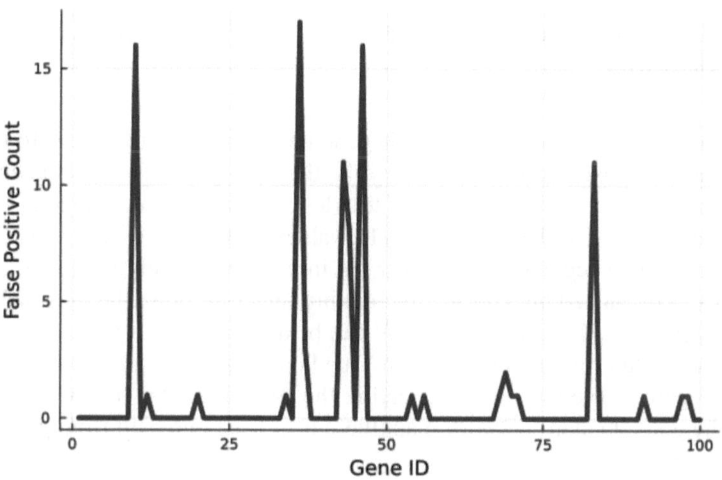

Fig. 3. Distribution of false positive connections per gene as the source, when selecting the top 200 inferred connections while the network contains 176 true connections. This figure highlights which genes are more prone to generating false positives in the regulatory network inference process

$W[i,:]$, respectively. By comparing against the gold standard, we observe that many false positive connections share common source genes. False positive connections for each gene as the source, denoted as FP_i, represent the number of false positive connections attributed to gene i as the source. As shown in Fig. 3, connections from a specific set of genes have higher false positive rates compared to others. Additionally, some rows in W have higher average values, indicating that these genes exert greater influence on other genes in the network. Consequently, ranking selection tends to prioritize connections with these influential genes as sources, as illustrated in Fig. 2.

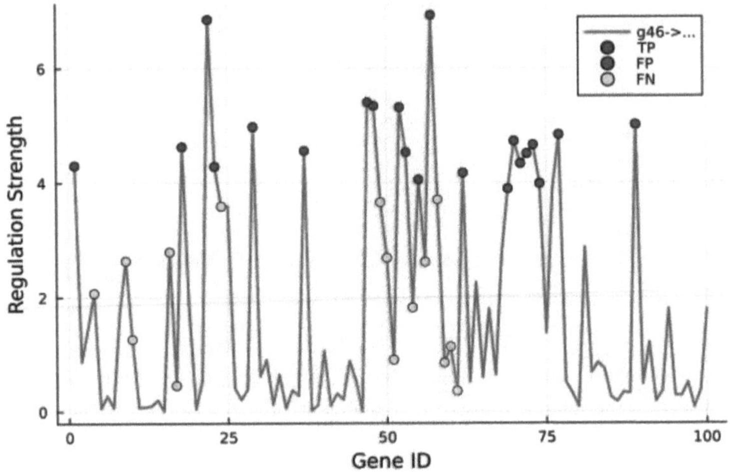

Fig. 4. Regulatory strength from gene 46 to other genes in the network, demonstrating the impact of indirect connections.

For example, the gold standard for this gene network shows that gene 46 is central: 24 true connections use g_{46} as the source while the average number of connections per gene is 7. The effect of turning off gene 46 trickles down into many indirectly connected genes via 1st-, 2nd-,... orders of edges. The values of $W_{46,i}$, $i = 4, 9, 10\ldots$, are high because of the direct regulation. However, an indirect path such as $g_{46} \rightarrow g_{37} \rightarrow g_1$ causes $W_{46,1}$ to be ranked high during the ranking selection step. The rank of **W** falsely selects many $g_{46} \rightarrow g_n$. A high rank of **W** can be a direct path or the consecutive path to genes downstream. In Fig. 4 we plot the True Positives (green), False Positives (red), and False Negatives (yellow) when selecting the 21 highest **W** values from gene 46. Based on the gold standard network map, there is at least one path from gene 46 to all false positive genes.

The connection probability $p(X_{i,j})$ from gene i to gene j is inversely proportional to the sum of **W**[:, j] from all genes to gene j, i.e., if many genes have high values $W_{k,j}$ on connections to gene j, some of the connections are potentially indirect and cause the increase of $W_{k,j}$ [1, 2].

$$p(\mathbf{X}_{i,j}) \propto W'[i, j] \tag{2}$$

$$W'[i, j] = W[i, j] / \sum_{m=1}^{n} W[m, j] \tag{3}$$

$$= W[i, j] / \left(\sum_{k \in \text{Indirect}[i,j]} W[i, k] + W[i, j] \right) \in gs[i, j] \tag{4}$$

To verify this hypothesis, we calculate the average **W** values between every pair of genes in the network of 5 cases: (1) no path between two genes, (2) only direct connections, (3) only indirect connections, (4) either, i.e., having direct or indirect connections, and (5) all pairs. Table 1 shows that **W** values between two genes with only a direct path are much higher than values between two genes with more than one path.

Table 1. Average **W** values

Network	No Path	Direct	Indirect	Either	All Pairs
N1	0.46	4.17	1.04	1.67	0.54
N2	0.55	3.22	0.81	1.36	0.61
N3	0.49	3.56	0.62	0.79	0.57
N4	0.51	3.2	0.7	0.88	0.6
N5	0.51	2.2	1.21	1.34	0.6

Indirect paths imply intermediate genes directly connecting to the sink. For example, $g_{46} \rightarrow g_{37} \rightarrow g_1$ implies that $W_{46,1}$ and $W_{37,1}$ are high values and $W_{46,1} < W_{37,1}$. Figure 5 shows that the True Positive connections have concentrated peaks around gene 46 while False Positives have multiple peaks spread around many genes.

We denote a gene that only appears once as the directly connected sink in the network as an *isolated sink.* For example, $g_{26} \rightarrow g_{35}$ is the only connection that has g_{35} as the sink, including both direct and indirect connections. g_{35} is an *isolated sink.*

In **Algorithm** 1 we describe how we calculate the normalized matrix $\mathbf{W}'$. Figure 6 plots the **W** values for some isolated sinks, along with $\mathbf{W}'$ values, showing a single large peak with low noise around the source gene.

Algorithm 1 Dom Norm
(Normalization for Gene Interactions)

Input: W matrix (a table of gene interactions)
Hyper-parameters: $\theta = 0.7$;
Output: W' matrix (normalized gene interactions)
for each gene j **do**
  Step 1: Thresholding (removing weak interactions)
  for each gene i **do**
    threshold $\leftarrow \theta \times W[i, j]$
    if $W[i, j] <$ threshold **then**
      $W[i, j] \leftarrow 0$
    end if
  end for
  Step 2: Calculating the total interaction strength
  total_sum $\leftarrow \sum_{i=1}^{100} W[i, j]$
  Step 3: Normalizing the remaining interactions
  for each gene i from 1 to 100 **do**
    if $W[i, j] > 0$ **then**
      $W'[i, j] \leftarrow W[i, j]/$total_sum
    end if
  end for
end for
return W'

After normalization, Fig. 7 shows that the **W** values of false positives (the mid-range regulation strength genes which are false positives on the right side of the graph) decrease more relative to the false negatives. Equation 3 shows that the regulation is further separated (into direct vs. indirect) if there is a path from gene *i* to gene *j*. The clustering method helps distinguish between direct and indirect connections by differentiating between low-expressed regulation strengths, which are direct connections, and false positive connections that have been influenced more after normalization. This is because, in Algorithm 1, the false positives, which correspond to mid-range regulatory strength and are mainly due to indirect connections, are normalized more significantly compared to normal genes. This results in previously identified false positives becoming more comparable to the lower regulation strengths of direct connections. Therefore, we apply the clustering method in Algorithm 2 to separate these newly mixed-up false positives from low-expressed, direct connections.

$$p(x_{i,j}) = \begin{cases} 1 & \text{if } W'[i, j] \geq \text{threshold} \\ W'[i, c1] \cdot \rho_1 & \rho_1 < 1, \\ W'[i, c2] \cdot \rho_2 & \rho_2 > 1. \end{cases} \tag{5}$$

Equation 5 estimates the probability of direct connections using a threshold, assuming that high regulation strengths indicate direct connections. In our experiments, we choose $\rho_1 = 0.5$ and $\rho_2 = 1.5$. The clusters $c1$ and $c2$ are obtained as described in Algorithm 2. With the results of Algorithm 2, we can then differentiate between false positives (FPs) and true positives (TPs). Two arguments support this approach:

1. Mid to low **W′** values are less likely to indicate direct connections.
2. After normalization, many indirect connections are reduced to sufficiently low values, making them overlap with direct connections.

The smaller cluster is more likely to represent direct connections, while the larger cluster indicates the absence of direct connections.

Algorithm 2 Cluster Norm

```
Input: W′ ▷ Matrix of normalized gene regulatory data after applying Dom Norm
Output: W″
for each gene i do
    Define the high and low bounds r1, r2 based on the maximum value in row W′[i, :];
    Select a subset of values within the bounds: Identify entries in W′[i, :] such that r1 ≤ W′[i, :] ≤ r2;
    Cluster the subset: Apply a clustering algorithm to the subset of W′[i, :] within the bounds to form two clusters: c1 and c2;
    Ensure that cluster c1 is more likely to contain false positives: Enforce the condition |c1| ≥ |c2|, swapping labels if needed;
    Apply scaling factors:
    W″[i, c1] = W′[i, c1] × ρ1, where ρ1 < 1;    ▷ Reduce impact of low regulatory values
    W″[i, c2] = W′[i, c2] × ρ2, where ρ2 > 1;    ▷ Increase impact of high regulatory values
end for
```

3.2 High and Low Regulation

The false negatives in Fig. 4 suggest that many direct connections exhibit low *W* values. Gene 51 and gene 61 can be studied in this context, as gene 46 connects to both while having relatively low *W* values compared to other connections using 46 as the source. Figure 8 demonstrates that the *W* values from all genes to g_{51} are unstable and small, exhibiting considerable fluctuation. Although genes 8 and 70 are the two most dominant regulators, neither possesses outgoing edges or paths to g_{51}. Consequently, using the relative ranking of either *W* [:, 51] or *W* [46,:] fails to accurately reflect the connection between genes 46 and 51. For gene 61, multiple peaks appear in Fig. 8 that do not align with the gold standard connections; the true connections are genes 37, 46, and 63, while the highest peak occurs at gene 44. According to the ***gs*** matrix, we observe that $g_{44} \rightarrow \ldots \rightarrow g_{37}$ and $g_{44} \rightarrow \ldots \rightarrow g_{63}$, while g_{44} does not directly influence g_{61} ($g_{44} \not\rightarrow g_{61}$). Turning off gene 44 results in changes to genes 37 and 63, and the combined effects from 37 to 61 and 63 to 61 surpass any single direct connection involving gene 44 or 63. In contrast, the impact of gene 46 on gene 61 remains relatively low.

Some direct gene interactions are inherently weak. Rather than altering the expression level of the target gene, these interactions may serve other roles, such as acting as catalysts for biochemical processes or functioning within more complex regulatory mechanisms.

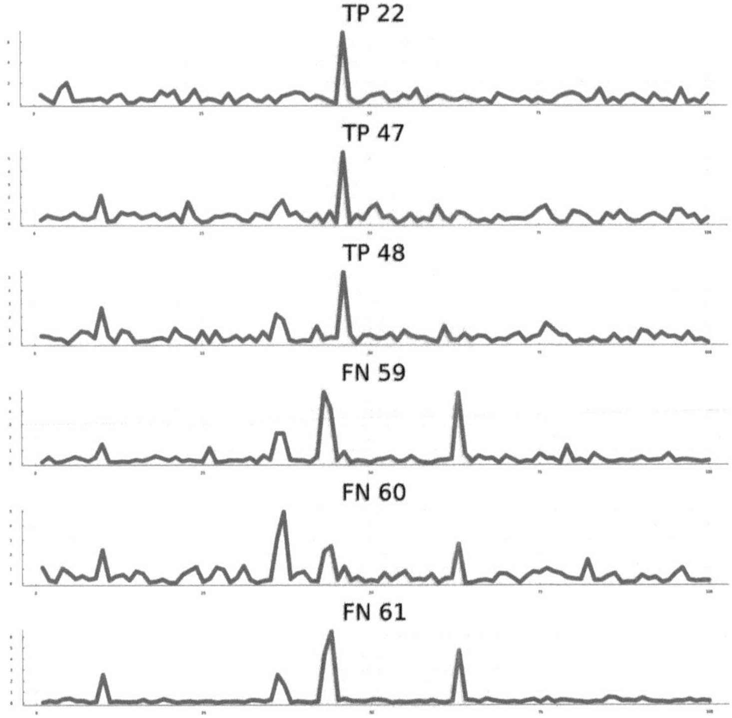

Fig. 5. Distribution of regulatory strength from gene 46 as observed in the matrix **W**[:, **46**]. The true positives (TPs) exhibit a single large peak centered around gene 46, indicating a strong regulatory influence. In contrast, the false positives (FPs) show multiple peaks across various genes, reflecting a tendency for incorrect regulatory connections inferred from the data.

4 Results

DREAM4 data [13] are available at https://gnw.sourceforge.net/dreamchallenge.html. DREAM3 data [12] are available at https://gnw.sourceforge.net/dreamchallenge.html. The method is implemented using Julia, available at https://github.com/ywang552/Gene.jl.

All experiments were conducted on a standard Windows PC with the following specifications: Intel(R) Core(TM) i7-9700 CPU @ 3.00 GHz, 32.0 GB RAM, 64-bit operating system. The additional runtime for Dom Norm and Cluster Norm is 0.05 s. The total runtime for one dataset of a 100-gene network is 0.2 s.

Both W' and W'' are derived as refined versions of the original W matrix from the MCZ methodology. To assess the effectiveness of Dom Norm, we evaluate five 100-gene networks from the DREAM3 and DREAM4 datasets, focusing on PR-AUC metrics. Results indicate that Dom Norm consistently outperforms the baseline MCZ approach, with notable PR-AUC improvements across all networks. For DREAM4, Network 1 exhibits a 17% increase in PR-AUC, while the overall average improvement across networks is 8.6%, as illustrated in Fig. 9. For DREAM3, the average improvement across all network is 6.8% as illustrated in Fig. 10.

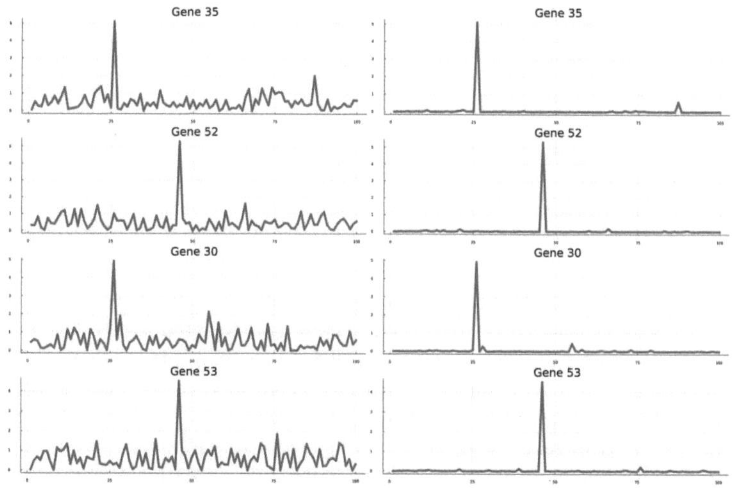

Fig. 6. Comparison of the weighted adjacency matrices for isolated sink genes before and after applying Dom Norm. The matrix W (before normalization) exhibits significant noise and fluctuations, while the normalized matrix W' reveals distinct singular peaks, indicating a more stable representation of direct regulatory connections.

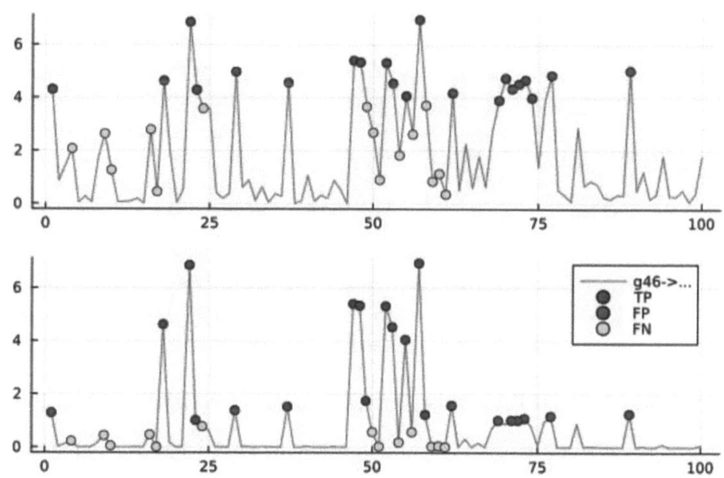

Fig. 7. Comparison of the W matrix before and after applying Dom Norm: TP increased from 9 to 11, FP reduced from 12 to 10, highlighting improved regulatory strength assessments and reduced indirect pathway influences.

For a more granular view, Precision-Recall curves for Network 1 (most straightforward to infer) and Network 5 (most challenging to infer) are presented in Fig. 11a and Fig. 11b, respectively. After applying **Dom Norm**, we observe a marked improvement in performance by attenuating the influence of indirect connections.

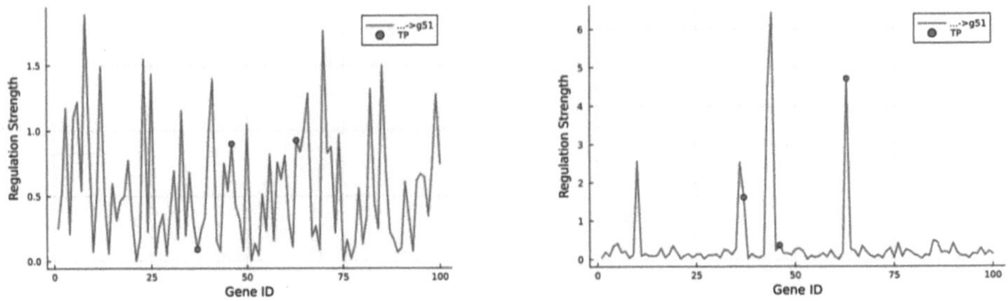

Fig. 8. Regulatory landscape of gene 46 towards its direct connections, genes g_{51} and g_{61}, as inferred from the weighted adjacency matrix **W**. According to the gold standard (gs), gene 46 has direct connections to both g_{51} and g_{61}. However, the values of W [46, 51] and W [46, 61] are relatively low compared to other connections where gene 46 serves as the source. This prompted an exploration of the landscapes of W [:, 51] and W [:, 61]. The plot for W [:, 51] reveals low and fluctuating values across all genes, suggesting minimal regulatory influence. In contrast, W [:, 61] displays several peaks, but these do not correspond to the true connection points, with the highest peak appearing at gene 44, which is not a direct regulator of g_{61}.

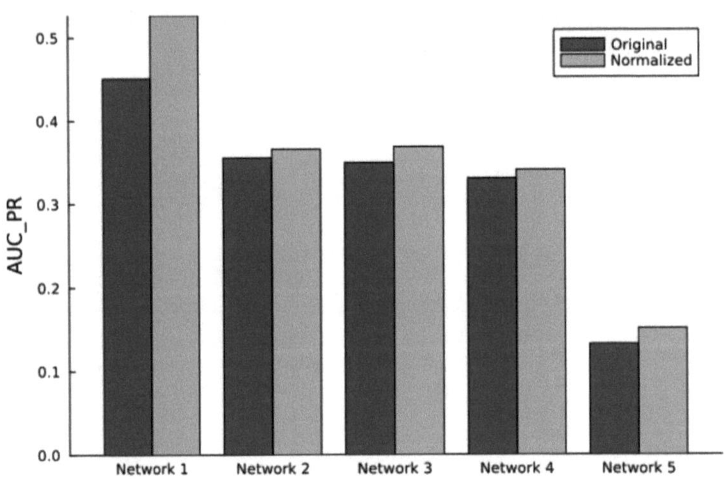

Fig. 9. Comparison of the DREAM4 dataset results using the Original MCZ (blue) and the Normalized MCZ (orange).

However, **Dom Norm** occasionally fails to adequately distinguish between indirect and weak direct connections. This is mitigated by **Cluster Norm**, which focuses on clusters of similar W values to further reduce false positives.

The precision-recall curves after applying Cluster Norm underscore its ability to more effectively separate these connections. In our experiments, Cluster Norm improves performance on Networks 1 and 5. As shown in Fig. 11, the precision-recall plot demonstrates that Cluster Norm (orange curve) consistently encompasses a larger area than Dom Norm (blue curve), highlighting its efficacy.

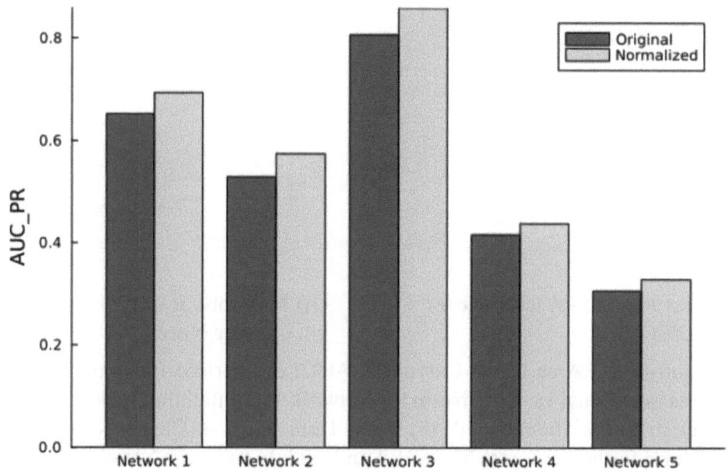

Fig. 10. Comparison of the DREAM3 dataset results using the Original MCZ (blue) and the Normalized MCZ (orange).

5 Conclusion

In this paper, we have proposed improvements to algorithms for gene regulation network inference using gene knockout data. The matrix of connection strengths (between gene pairs) is first modified by thresholding and normalization with respect to other connection strengths, reducing the impact of indirect connections. This is followed by a cluster normalization step that distinguishes low magnitude direct connections from indirect connections. Our approach combines ranking selection, Gaussian process probability modeling, and thresholding, thereby enhancing the quality of results obtained with gene regulation network inference and minimizing false positives.

The main novel contribution of this work is in differentiating between regulation strength estimates due to direct vs. indirect connections. Our unique approach provides an effective way to distinguish between these two types of connections, thereby improving the inference process. To validate the effectiveness of our approach, we achieved an average improvement of 8.6% in Area Under the Precision-Recall Curve (AUPC) scores, with a maximum improvement of 17% observed for the DREAM4 dataset. Additionally, for the DREAM3 dataset, our method demonstrated an average improvement of 6.8% Our method showed consistent improvements across all evaluated datasets, providing strong evidence for its utility in gene regulatory network inference. The DREAM5 dataset is excluded due to insufficient steady-state data. For example, Network 1 contains 1643 genes but only 400 knockout data points, with similar ratios observed across other networks.

Future work will focus on refining and expanding our current approach. First, we aim to improve the clustering algorithm by incorporating kernel functions, enhancing the distinction between true positives (TPs) and false positives (FPs) for more robust results. Second, we plan to explore the max-flow inference method [10, 14], with the expectation that refining this technique will lead to more accurate models of gene regulation. A

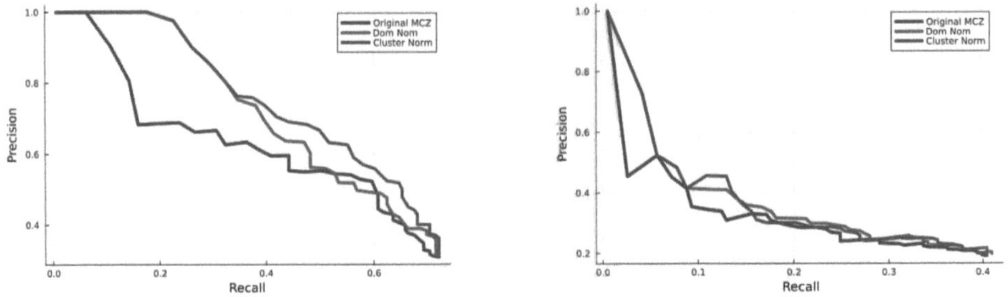

(a) Network 1: Original vs. Dom Norm vs. Cluster Norm

(b) Network 5: Original vs. Dom Norm vs. Cluster Norm

Fig. 11. Precision-Recall Area Under Curve (PR-AUC) comparison for Networks 1 and 5. The left side (a) displays Original vs. Dom Norm for Network 1, highlighting improvements in AUPC after applying Dom Norm. The right side (b) shows Dom Norm vs. Cluster Norm for Network 5, where Cluster Norm further improves Network 1 and slightly enhances Network 5, demonstrating its effectiveness in reducing false positives by distinguishing similar W values.

Gaussian process model can represent time series data as nodes and their regulation [6], while addressing indirect connections that may overshadow direct ones. We will develop more sophisticated methods to differentiate between these connections, ensuring that significant direct relationships are preserved in our network inference framework. This improvement will further enhance the precision and reliability of our gene regulatory network models.

References

1. Aalto, A., Viitasaari, L., Ilmonen, P., Mombaerts, L., Goncalves, J.: Continuous time Gaussian process dynamical models in gene regulatory network inference. arXiv preprint arXiv:1808.08161 (2018)
2. Aalto, A., Viitasaari, L., Ilmonen, P., Mombaerts, L., Gonçalves, J.: Gene regulatory network inference from sparsely sampled noisy data. Nat. Commun. **11**(1), 3493 (2020)
3. Al-Mubaid, H., Aldwairi, T.: Utilizing functional annotation of disease genes for disease clustering. In: Proceedings of International Conference on Bioinformatics, vol. 92, pp. 58–71 (2023)
4. Al-Mubaid, H., Nguyen, H.A.: A cluster-based approach for semantic similarity in the biomedical domain. In: 2006 International Conference of the IEEE Engineering in Medicine and Biology Society, pp. 2713–2717. IEEE (2006)
5. Bravo-Hermsdorff, G., Watson, D., Yu, J., Zeitler, J., Silva, R.: Intervention generalization: a view from factor graph models. Adv. Neural Inf. Process. Syst. **36** (2024)
6. Cui, C., Banelli, P., Djurić, P.M.: Topology inference of directed graphs by Gaussian processes with sparsity constraints. IEEE Trans. Sig. Process. (2024)
7. Gamage, H.N., Chetty, M., Lim, S., Hallinan, J.: GRAMP: a gene ranking and model prioritisation framework for building consensus genetic networks. Knowl.-Based Syst. **302**, 112374 (2024)
8. Greenfield, A., Madar, A., Ostrer, H., Bonneau, R.: DREAM4: combining genetic and dynamic information to identify biological networks and dynamical models. PLoS ONE **5**(10), e13397 (2010)

9. He, X., Wang, Y., Jin, J.: Bayesian inference and optimisation of stochastic dynamical networks. Int. J. Syst. Sci. 1–15 (2024)
10. Kim, K., Lee, D., Essa, I.: Gaussian process regression flow for analysis of motion trajectories. In: 2011 International Conference on Computer Vision, pp. 1164–1171. IEEE (2011)
11. Li, X., Ma, Y., Zhou, Q., Zhang, X.: Sparse large-scale high-order fuzzy cognitive maps guided by spearman correlation coefficient. Appl. Soft Comput. **167**, 112253 (2024)
12. Madar, A., Greenfield, A., Vanden-Eijnden, E., Bonneau, R.: DREAM3: network inference using dynamic context likelihood of relatedness and the inferelator. PLoS ONE **5**(3), e9803 (2010)
13. Marbach, D., Prill, R.J., Schaffter, T., Mattiussi, C., Floreano, D., Stolovitzky, G.: Revealing strengths and weaknesses of methods for gene network inference. Proc. Natl. Acad. Sci. **107**(14), 6286–6291 (2010)
14. Pareek, P., Nguyen, H.D.: A framework for analytical power flow solution using Gaussian process learning. IEEE Trans. Sustain. Energy **13**(1), 452–463 (2021)
15. Wang, Y., Mohan, C.K.: Gene regulatory network inference with evolution strategies and sparse matrix representation. In: 2019 IEEE International Conference on Bioinformatics and Biomedicine (BIBM), pp. 2105–2112. IEEE (2019)
16. Wehe, A., Bansal, M.S., Burleigh, J.G., Eulenstein, O.: DupTree: a program for large-scale phylogenetic analyses using gene tree parsimony. Bioinformatics **24**(13), 1540–1541 (2008)
17. Wei, P.J., et al.: Inference of gene regulatory networks based on directed graph convolutional networks. Brief. Bioinform. **25**(4), bbae309 (2024)

Adverse Effect Analysis for Psychiatric Medication Through In-Context Pruning Framework

S. Vismaya[1(✉)], Khushi Jayakumar Patil[1], B. V. Varun[1], Arun Amar Kurali[1], and Bhaskarjyoti Das[2]

[1] Department of CSE, PES University, Bangalore, India
12vismaya@gmail.com
[2] Departmenrt of CSE (AI & ML), PES University, Bangalore, India

Abstract. Adverse drug events (ADEs) contribute significantly to hospital admissions and increased mortality rates. Early recognition of these reactions remains crucial. Traditional systems like the FDA Adverse Event Reporting System (FAERS) have been key to drug safety surveillance, but social media platforms like Reddit now offer a rich source of real-world insights, transforming pharmacovigilance. To tackle the challenge of extracting meaningful insights from vast, unstructured data, Large Language Models (LLMs) offer an invaluable solution. In this paper, we present "In-Context Pruning Framework" (ICPF), a novel methodology that utilizes a fine-tuned LLaMA-3 model to filter irrelevant content and focus on context-sensitive ADE analysis. Our approach combines SciBERT for entity extraction, and synonym normalization to enhance data consistency. We demonstrate the applicability of this methodology in the domain of Psychiatry by identifying ADRs associated with Psychiatric drugs. Additionally, we conduct co-occurrence and drug class-level analysis to uncover intricate relationships between drugs and their adverse effects. Discrepancy between the ADE extracted from Reddit and the official source is also being presented thereby bridging the gap between user-reported experiences and official sources, providing actionable insights to improve pharmacovigilance and patient safety.

Keywords: Adverse drug events (ADEs) · Social Media · Pharmacovigilance · Large Language Models (LLMs) · In-Context Pruning Framework (ICPF) · Psychiatric Drugs

1 Introduction

Adverse Drug Reactions (ADRs) pose a significant challenge in healthcare, with underreporting compromising effective pharmacovigilance. In India, for instance, the ADR reporting rate is alarmingly low at under 1%, compared to a global average of 5% [1]. Factors contributing to this gap include limited awareness among healthcare professionals, inefficient reporting systems, heavy patient loads, and widespread self-medication practices [2]. A systematic review found that ADRs account for 2.85% of hospital admissions,

T. Aldwairi et al. (Eds.): BICOB 2025, CCIS 2535, pp. 42–54, 2025.
https://doi.org/10.1007/978-3-031-94039-2_4

with 6.34% of hospitalized patients developing ADRs [3]. Vulnerable populations—such as the elderly, women, and individuals on multiple medications are disproportionately affected. Addressing these challenges, particularly in underserved regions, requires innovative approaches such as raising awareness, adopting telemedicine, and leveraging AI and social media data [1].

Effective ADR monitoring necessitates both pre and post marketing surveillance. Pre-marketing clinical trials, while essential, are often constrained by small sample sizes and limited follow-up periods, creating gaps in the understanding of rare or long-term adverse effects. Comprehensive post-marketing surveillance bridges these gaps, providing crucial real-world insights into medication safety.

Globally, ADRs are a major public health concern, contributing to 5.9%–22.3% of emergency cases and approximately 106,000 deaths annually in the United States alone [34]. Historical incidents like the thalidomide disaster of the 1960s, which caused severe birth defects, underscore the critical need for robust pharmacovigilance systems [4]. Delayed recognition of adverse effects has been evident in cases like aspirin's gastrointestinal risks and phenacetin's renal toxicity. Psychiatric medications have also faced scrutiny, with older drugs causing life-threatening arrhythmias and newer ones, such as SSRIs, linked to severe side effects [23, 24]. For example, clozapine, a life-saving medication for treatment-resistant schizophrenia, carries a significant risk of myocarditis-related mortality [25]. These examples highlight the limitations of clinical trials in identifying rare adverse effects, emphasizing the need for ongoing monitoring and flexible prescribing practices to ensure patient safety.

2 The Role of Public Forums in Pharmacovigilance

Traditional pharmacovigilance systems, such as the FDA's Adverse Event Reporting System (FAERS), are crucial but often miss key patient experiences, particularly for psychiatric medications, due to underreporting. Public forums like Reddit and health-focused discussion boards offer real-time, diverse insights from a broad patient base, capturing Adverse Drug Events (ADEs) that may not appear in official databases. These platforms are valuable for identifying under-reported side effects, such as sexual dysfunction from antidepressants, which is frequently overlooked and leads to treatment non-adherence [36], as well as long-term effects like cognitive impairments or mood changes associated with SSRIs [37]. Despite their potential, public forums present challenges due to unstructured and noisy data, requiring advanced natural language processing techniques for extracting reliable, medically relevant information [38].

Recent studies show that social media platforms enhance pharmacovigilance, with large-scale analyses of platforms like Twitter, revealing antidepressant side effects such as sleep disturbances and weight changes [5], and advanced algorithms like Word2Vec improving ADE detection [7]. Comparisons between social media-derived ADR data and FAERS have highlighted overlaps, uncovering previously unreported side effects [8]. Additionally, public forums help identify discrepancies between patient-reported side effects and those listed in regulatory documents, providing regulators with a more comprehensive understanding of a drug's impact [10]. Integrating insights from these forums into pharmacovigilance frameworks can enhance drug safety monitoring and improve patient outcomes [6].

3 Leveraging LLMs for Pharmacovigilance

Recent advancements in artificial intelligence (AI) and natural language processing (NLP) have significantly improved pharmacovigilance, particularly in detecting Adverse Drug Events (ADEs). Traditional methods, reliant on manual reporting, are time-consuming and often lead to underreporting. In contrast, large language models (LLMs) such as LLaMA have demonstrated their ability to efficiently process and classify unstructured data from social media, improving the accuracy of ADE detection. Early research has used NLP models to classify ADR-related posts from social media, achieving high precision in automated ADE monitoring [11, 12]. Recent developments have integrated fact-checking mechanisms and showcased LLMs' utility in assessing medication side effects, like those from ADHD medications, with exceptional accuracy [13, 14].

4 Related Work

Recent research has explored various methods for detecting Adverse Drug Reactions (ADRs) from social media, leveraging the unstructured nature of user-generated content to enhance pharmacovigilance. The ADR Detection Framework (ADF) [19] used NLP techniques to extract ADRs from health forums, focusing on diabetes drug reviews. However, it struggles with informal language and lacks access to clinical data, limiting its ability to make causal inferences. ADRMine, a machine learning-based system, used Conditional Random Fields and Word2Vec to detect ADRs from social media posts, achieving good performance on DailyStrength and Twitter data, but it faces challenges with ambiguous sentences and feature engineering [20].

A causality-sensitive method proposed by [21] improved ADR detection by identifying causal relationships but relied on manual annotations and lacked integration with medical ontologies, limiting its scalability. Additionally, data augmentation using Variational Autoencoders (VAEs) [22] reduced the need for annotated data but struggled with sentence complexity and generalization across platforms. These studies highlight the promise of social media for ADR detection while underscoring the need for better handling of informal language, integration of deep learning models, and scalable frameworks across diverse platforms.

5 Methodology

In this paper, we introduce a novel methodology, "In-Context Pruning" (ICP), which leverages large language model (LLM) capabilities, specifically employing a fine-tuned LLaMA model to perform adverse drug effect (ADE) classification from unstructured social media text. Recent studies have demonstrated LLaMA-2's superiority in text classification tasks, with the 13B and 70B versions surpassing GPT-3 in efficiency and effectiveness [15].The ICP framework tackles three primary limitations commonly found in traditional supervised machine learning approaches:

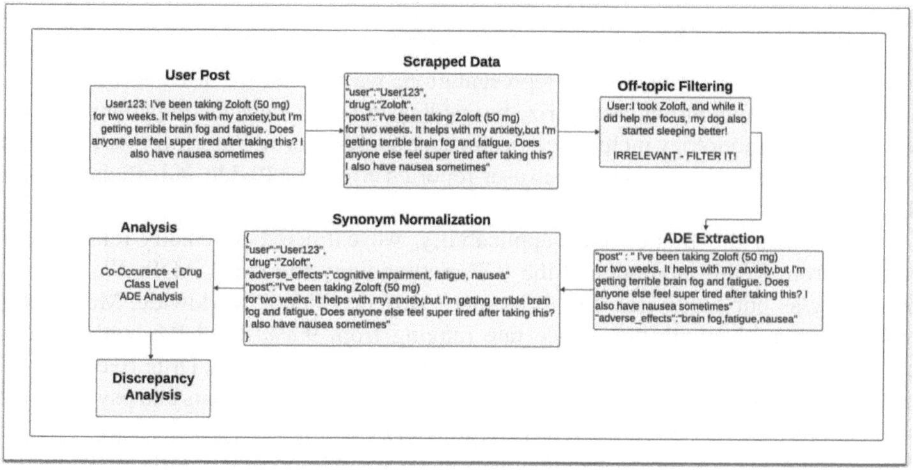

Fig. 1. ICP Framework

Unlike static word embeddings like Word2Vec, which provide a fixed representation for each word regardless of context, our fine-tuned LLaMA model captures the full contextual meaning of words in sentences. This context-aware understanding is critical in processing informal and unstructured text found on social media platforms, where ADEs may be described in nuanced and varied ways. By fine-tuning LLaMA, the ICP method can discern subtle distinctions in language and map colloquial expressions to standardized medical terms, significantly improving the accuracy and reliability of ADE detection.

Traditional supervised machine learning approaches often require large amounts of annotated data, which can be costly and challenging to obtain. While our approach involves fine-tuning, the use of LLaMA's pre-trained capabilities allows the model to generalize effectively even with a limited number of labeled examples, reducing the dependency on extensive annotated datasets. This enables our framework to operate efficiently across multiple health domains.

Supervised models typically struggle to identify ADEs that were not explicitly present in their training data, limiting their capacity to adapt to new or underrepresented health domains. Our fine-tuned LLaMA model's rich pre-trained knowledge base enables it to capture diverse and previously unreported ADEs, overcoming this limitation. This adaptability makes our ICP approach scalable and effective for analyzing ADEs across various medical conditions, including psychiatric medications and other complex health scenarios.

The ICP methodology involves several key components that work together to effectively detect Adverse Drug Reactions (ADRs) from unstructured social media data. First, we collected and preprocessed Reddit posts related to six psychiatric conditions, anonymizing user data for privacy and organizing it in JSON format for efficient processing. We then fine-tuned the quantized LLaMA-3 model for ADE classification, using parameter-efficient techniques such as Low-Rank Adaptation (LoRA) and mitigated class imbalance through weighted loss functions. To identify drug names and associated ADRs, we applied a SciBERT-based model for Named Entity Recognition (NER). For consistency in medical term representation, we utilized the MedNorm database to normalize synonyms and categorize biomedical concepts detected in patient reviews. Finally, our methodology includes co-occurrence analysis, drug class-level studies, and a comparison of discrepancies between user-reported effects on Reddit and official FAERS data.

To evaluate the framework's applicability, we extracted psychiatry-related drug reviews from Reddit and applied the ICP methodology to extract ADRs.We selected psychiatry as our focus area due to its increasing prevalence worldwide. Mental disorders have a 12-month prevalence rate ranging from 9.8% to 19.1%, with lifetime prevalence between 18.1% and 36.1% [16]. Studies report Adverse Drug Event (ADE) incidence rates of 24% to 38.2% among patients using antidepressants and psychotropic medications [17, 18]. As most ADRs are mild and classified as "possible" with limited preventability [17, 18], detecting previously unknown or unreported ADRs in psychiatric drugs is critical for enhancing patient safety and improving treatment outcomes. We then conducted a comparative analysis of ADRs identified across six medical conditions to uncover valuable insights. Finally, we validated our ICP framework by benchmarking it against the FAERS database, demonstrating its effectiveness in uncovering previously unreported ADRs.

Our contributions are threefold. First, we developed and implemented the ICP framework, which can identify ADRs in online drug reviews shared by patients across various health domains, including mental health. Second, we showcased the effectiveness of the ICP framework by applying it to drug reviews related to mental health. This process enabled the automatic identification, analysis, and visualization of ADRs, revealing valuable insights such as common and rare ADRs, co-occurrence patterns of ADEs, and previously unknown ADRs, verified through benchmarking against the FAERS database. These findings are impactful for healthcare professionals, regulatory bodies, and pharmaceutical companies, providing new perspectives on the adverse effects of psychiatric medications. Finally, the framework extends beyond ADR detection, with potential applications in other health domains and integration with electronic health records, supporting improved data-driven decision-making. Additionally, our work provides a foundation for future research on real-time drug safety surveillance leveraging social media and other unstructured data sources.

5.1 Data Collection and Preprocessing

Reddit posts were collected to analyze mentions of Adverse Drug Effects (ADEs) related to medications commonly prescribed for mental health conditions, including ADHD, anxiety, insomnia, depression, bipolar disorder, and schizophrenia. We considered 5 drugs for each medical conditions and almost 100,000 posts were scraped about each drug. Posts and comments were scraped in JSON format (as shown in Fig. 1), with each entry containing information about the post, medication name, and author details, which were anonymized with unique IDs for privacy. The FDA Adverse Event Reporting System (FAERS), containing nearly 800,000 annual reports [26], was used as a benchmark to validate our findings.

5.2 Llama Model Fine-Tuning for ADE Classification

Here, we fine-tuned a quantized LLaMA-3 model for sequence classification on the ADE Corpus V2 dataset, specifically the "Ade corpus v2 classification" subset.

Our approach includes dataset preparation, model quantization, and customization of the training process with a weighted loss function to handle class imbalance. As detailed in Table 1, a total of 4,411 public user reviews spanning 30 different psychiatric drugs were filtered across six medical conditions after the classification process.

Table 1. Medical Conditions, Drug Counts, and Total Reviews

Medical Condition	Drug Count	Total Reviews
ADHD	5	703
Anxiety	5	909
Bipolar	5	819
Depression	5	731
Insomnia	5	1100
Schizophrenia	5	149

For our base model, we used LLaMA-3 with 8 billion parameters (meta-llama/Meta-Llama-3-8B). The model was configured with BitsAndBytesConfig to enable 4-bit quantization, which significantly reduces memory usage and allows for efficient processing on hardware with limited GPU memory.

We applied Low-Rank Adaptation (LoRA) for efficient parameter fine-tuning, using a configuration with a rank of 16, a scaling factor of 8 for the update matrix, and targeting specific modules ("q proj", "k proj", "v proj", "o proj") for adaptation. To reduce overfitting, we set a LoRA dropout of 0.05. This LoRA configuration enabled k-bit training, allowing the model to perform the classification task without fine-tuning all parameters.

To address class imbalance, a custom Trainer class was implemented, extending Hugging Face's Trainer to incorporate class weights in the loss calculation. If class weights were specified, cross-entropy loss was computed with these weights; otherwise, standard cross-entropy loss was used. This approach combines quantization, LoRA fine-tuning, and class-weighted loss to create a memory-efficient, class-balanced sequence classification model for detecting adverse drug effects. While the base Meta-LLaMA-3-8B model achieved 51.33% accuracy, our fine-tuned model reached 95% accuracy (Fig. 2).

```
Accuracy: 0.9500
Precision: 0.9435
Recall: 0.9573
F1 Score: 0.9504
Classification Report:
              precision    recall  f1-score   support

           0       0.96      0.94      0.95       750
           1       0.94      0.96      0.95       750

    accuracy                           0.95      1500
   macro avg       0.95      0.95      0.95      1500
weighted avg       0.95      0.95      0.95      1500

Balanced Accuracy Score: 0.95
Accuracy Score: 0.95
```

Fig. 2. Llama 3 fine-tuned model

5.3 ADE Extraction

To extract Adverse Drug Events (ADEs) from classified Reddit posts, we fine-tuned a BERT model for Named Entity Recognition (NER) using a custom token-labeling strategy. We used the "Ade corpus v2 drug ade relation" subset, which labels drug names and their associated adverse effects.

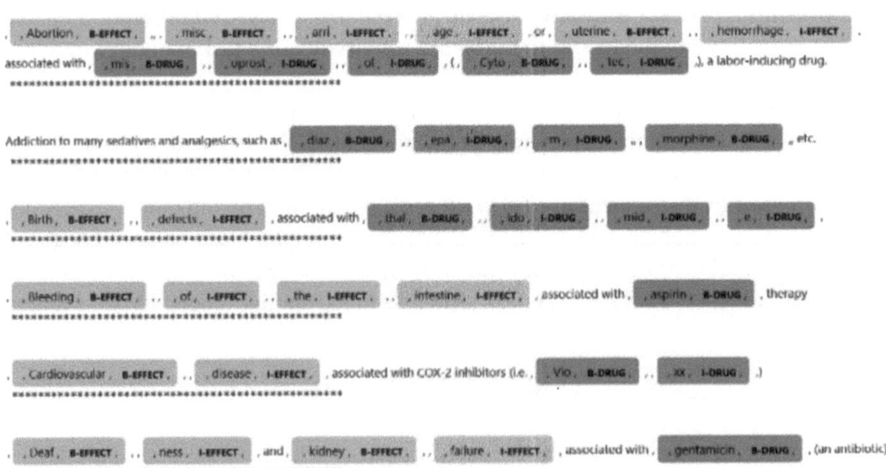

Fig. 3. Output visualisation

Two entities were targeted: DRUG (medication) and EFFECT (adverse effect). The data was processed with a BIO tagging scheme (Begin-Inside-Outside) to mark entities, yielding 5 possible labels: O (outside), B-DRUG, I-DRUG, B-EFFECT, and I-EFFECT. The tokenization process involved mapping each token to its corresponding BIO tag based on its position in the entity. After fine-tuning, the model accurately extracted ADE-related entities from Reddit posts, facilitating a better understanding of drug-effect relationships and providing valuable pharmacovigilance insights. The model the results are visualized in Fig. 3.

5.4 Synonym Normalization

To accurately determine the prevalence rate of each ADR (Adverse Drug Reaction), we standardize ADR names to align with their official definitions. User-generated content often contains informal language, slang or non-standard terminologies to describe adverse effects. To standardize these terms and map them to medically recognized terminologies, we employed a synonym normalization approach utilizing the MedNorm database. Our method involves using a Sentence Transformer model to measure the semantic similarity between the extracted Adverse Drug Event (ADE) terms from user content and ADE terms in the MedNorm database. By calculating similarity scores, we identify the most semantically similar term from MedNorm and map it to its corresponding MedDRA (Medical Dictionary for Regulatory Activities) code. For example, terms like "brain fog" are matched and normalized to "cognitive impairment", while "tired" is normalized to "fatigue", ensuring a consistent and medically accurate representation of adverse effects. Table 2 shows the mapping of extracted ADE to standard terms.

Table 2. MedDRA Terms and Extracted ADEs

MedDRA Code	MedDRA Term	Extracted_ade
10021111	Impulsivity	Lack of impulse control
10039077	Restlessness	Difficulty sitting still
10025498	Mania	Hyperactive mood
10002433	Anhedonia	Loss of interest
10040891	Difficulty initiating sleep	Trouble falling asleep
10019660	Hallucinations	Seeing unreal objects

6 ADE Co-occurrence Analysis

Our co-occurrence analysis of adverse drug effects (ADEs) from public forums examined patterns of simultaneous mentions across psychiatric medications, revealing key relationships. For conditions such as Bipolar Disorder, Anxiety, ADHD, Depression, Schizophrenia, and Insomnia, we identified both condition-specific and broad symptom patterns.

For example, Bipolar Disorder showed associations between Oropharyngeal pain, headache and parosmia, while Anxiety linked somnolence to fatigue as shown in Fig. 4. ADHD revealed connections between insomnia and attention deficit, and Depression displayed strong ties between affective symptoms and physical manifestations like panic attack. Schizophrenia exhibited a fragmented symptom network, with psychotic features tightly connected, and Insomnia highlighted the impact of sleep issues on daily functioning. At the domain level, insomnia and cognitive dysfunction were common across disorders, emphasizing the need for a tailored, nuanced approach to psychiatric medication management. These findings support personalized treatment plans to minimize adverse effects and optimize patient outcomes.

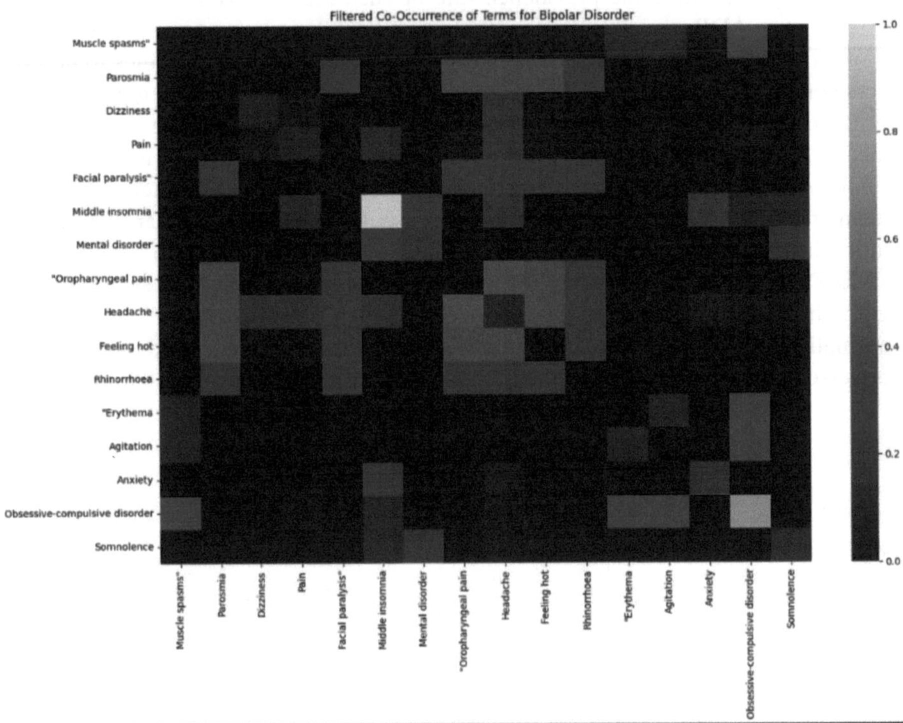

Fig. 4. Bipolar-Disorder Co-occurrence

7 Drug Class Level ADE Analysis

The analysis reveals critical findings on Adverse Drug Events (ADEs) across various conditions, with anxiety and insomnia being the most frequently discussed on Reddit, reflecting their widespread prevalence and treatment challenges. Ambien and zoloft were the drugs on which ADEs were most frequently reported highlighting the widespread use of insomnia and depression related drugs. Schizophrenia discussions highlighted ADEs such as insomnia and anxiety, while bipolar disorder posts often emphasized weight gain and tremor, pointing to the metabolic and neurological burdens of treatment.

For ADHD, insomnia and hyperactivity were the most common ADEs, appearing in over 28% of mentions, reflecting the impact of stimulant medications. These findings emphasize the shared prevalence of ADEs like insomnia and anxiety across conditions and underscore the importance of personalized treatment strategies to improve medication adherence, address metabolic side effects, and enhance overall quality of life for patients.

8 Discrepancy Analysis

For each drug associated with mental health conditions like ADHD, anxiety, insomnia, depression, and bipolar disorder, we identified ADEs that were frequently reported on Reddit but not listed in FAERS. This analysis reveals significant real-world insights and underscores the potential of social media data to capture underreported ADEs. For instance, users on Reddit associated Depression medications like Zoloft, Ketamine and Celexa with symptoms such as "chest pain" and "Sexual dysfunction"—ADEs that are absent from FAERS reports. Similarly, anxiety medications (e.g., Paxil) were linked with "Increased appetite" and "OCD" and insomnia medications (e.g., Ambien) with "memory impairment" and "sensory loss". These discrepancies may arise due to FAERS's dependence on clinical reporting, which might underrepresent user experienced symptoms, particularly for psychiatric medications. Social media mining offers a valuable supplement to traditional pharmacovigilance. It captures a broader range of patient experiences and potential ADEs that may warrant further clinical investigation.

9 Conclusion

In this study, we introduced the "In-Context Pruning Framework" (ICPF), leveraging advanced language models to improve the detection of Adverse Drug Events (ADEs) in user-generated content on public forums. By fine-tuning a quantized LLaMA model and integrating entity recognition through SciBERT, we successfully extracted meaningful insights from unstructured data sources like Reddit. Our findings highlight critical discrepancies between user-reported experiences and official sources, emphasizing the need for a complementary approach to traditional pharmacovigilance methods.

The results demonstrated that public forums are a rich yet underutilized resource for monitoring psychiatric medication effects. Through co-occurrence and drug class-level analyses, we identified not only common but also previously unreported ADEs, providing new perspectives on drug safety. This framework offers a scalable and efficient tool for pharmacovigilance, paving the way for enhanced patient safety and more informed clinical decisions.

10 Limitations

Despite the novel contributions and potential applications of the In-Context Pruning Framework (ICPF) for adverse drug reaction (ADR) analysis, several limitations must be acknowledged. The framework's reliance on ADRs extracted from online patient-based drug reviews raises concerns about the validity of these reports, as patient accounts on social media are often anecdotal and lack clinical validation. Misattribution of ADRs is a significant challenge, particularly for patients with co-morbidities or those taking multiple medications (polypharmacy), as they may incorrectly attribute an ADR to a specific drug, complicating causal assessment. Effective causality assessment often requires personal health records, including medical history, laboratory results, or details of co-administered medications—data not available on social media platforms. Furthermore, polypharmacy, common among psychiatric patients, complicates the detection of ADRs unique to a single medication, as drug-drug interactions are a significant source of adverse effects but are challenging to isolate using this framework.

11 Future Work

Future work could expand the dataset by including a broader range of public forums social media platforms, and health discussion boards. Integrating real-world data sources like electronic health records (EHRs) could further enhance the system's robustness. Addressing the complexities of polypharmacy, future studies could focus on disentangling Adverse Drug Reactions (ADRs) arising from drug-drug interactions, particularly in psychiatric treatments where multiple medications are commonly prescribed. Additionally, an evolution analysis of adverse effects over time could track how side effects emerge, persist, or diminish, providing insights into drug safety trends and correlations with clinical guidelines or public health interventions. The methodology could also be extended to other healthcare areas, such as chronic disease management or vaccine safety, to validate its versatility and scalability.

References

1. Shukla, S., et al.: Current scenario and future prospects of Adverse Drug Reactions (ADRs) monitoring and reporting mechanisms in the rural areas of India. Curr. Drug Saf. **19**(2), 172–190 (2024). https://doi.org/10.2174/1574886318666230428144120
2. Dutta, A., et al.: Analysis of under-reporting of adverse drug reaction: scenario in India and neighbouring countries. IP Int. J. Compr. Adv. Pharmacol. **5**(3), 118–124 (2020)
3. Patel, T.K., Patel, P.B.: Incidence of adverse drug reactions in Indian hospitals: a systematic review of prospective studies. Curr. Drug Saf. **11**(2), 128–136 (2016)
4. Vilhelmsson, A.: Consumer narratives in ADR reporting: an important aspect of public health? Experiences from reports to a Swedish consumer organization. Front. Public Health **3**, 211 (2015)
5. Saha, K., et al.: Understanding side effects of antidepressants: large-scale longitudinal study on social media data. JMIR Ment. Health **8**(3), e26589 (2021)
6. Toosi, F.G., Toussi, A.G.: Post-development sentiment and statistical analysis of different groups of psychiatric medications. Eur. J. Transl. Clin. Med. **7**, 47–56 (2024)
7. Nguyen, T., et al.: Estimation of the prevalence of adverse drug reactions from social media. Int. J. Med. Inform. **102**, 130–137 (2017)
8. Smith, K., et al.: Methods to compare adverse events in twitter to FAERS, drug information databases, and systematic reviews: proof of concept with adalimumab. Drug Saf. **41**, 1397–1410 (2018)

9. Farooq, H., et al.: Leveraging digital media data for pharmacovigilance. In: AMIA Annual Symposium Proceedings, vol. 2020 (2021)
10. Zhou, Z., Hultgren, K.E.: Complementing the US food and drug administration adverse event reporting system with adverse drug reaction reporting from social media: comparative analysis. JMIR Public Health Surveill. **6**(3), e19266 (2020)
11. Yang, M., Kiang, M., Shang, W.: Filtering big data from social media–building an early warning system for adverse drug reactions. J. Biomed. Inform. **54**, 230–240 (2015)
12. Bomgni, A.B., et al.: NLPADADE: leveraging natural language processing for automated detection of adverse drug effects. In: 2023 IEEE International Conference on Bioinformatics and Biomedicine (BIBM). IEEE (2023)
13. Prakash, S., Venkata, M., et al.: Accelerating pharmacovigilance using large language models. In: Proceedings of the 17th ACM International Conference on Web Search and Data Mining (2024)
14. Bannett, Y., et al.: Leveraging a Large Language Model to Assess Quality-of- Care: Monitoring ADHD Medication Side Effects. medRxiv (2024)
15. Kumar, A., Sharma, R., Bedi, P.: Towards optimal NLP solutions: analyzing GPT and LLaMA-2 models across model scale, dataset size, and task diversity. Eng. Technol. Appl. Sci. Res. **14**(3), 14219–14224 (2024)
16. Kessler, R.C., et al.: The global burden of mental disorders: an update from the WHO World Mental Health (WMH) surveys. Epidemiol. Psychiatr. Sci. **18**(1), 23–33 (2009)
17. Das, P., Nayak, J., Swain, S.P.: Adverse drug reaction monitoring of antidepressant drugs in a mental health institute in Odisha. Res. J. Pharm. Technol. **14**(12), 6479–6483 (2021)
18. Jayasree, A., et al.: Assessment of adverse drug reactions in treating common mental health disorders-findings from a tertiary care teaching hospital in Puducherry. J. Young Pharm. **16**(2), 322–327 (2024).
19. Oyebode, O., Orji, R.: Identifying adverse drug reactions from patient reviews on social media using natural language processing. Health Inform. J. **29**(1), 14604582221136712 (2023)
20. Nikfarjam, A., et al.: Pharmacovigilance from social media: mining adverse drug reaction mentions using sequence labeling with word embedding cluster features. J. Am. Med. Inform. Assoc. **22**(3), 671–681 (2015)
21. Bollegala, D., et al.: Causality patterns for detecting adverse drug reactions from social media: text mining approach. JMIR Public Health Surveill. **4**(2), e8214 (2018)
22. Mesbah, S., et al.: Training data augmentation for detecting adverse drug reactions in user-generated content. In: Conference on Empirical Methods in Natural Language Processing (2019)
23. Pacher, P., Kecskemeti, V.: Cardiovascular side effects of new antidepressants and antipsychotics: new drugs, old concerns? Curr. Pharm. Des.. Pharm. Des. **10**(20), 2463–2475 (2004)
24. Manolis, T.A., Manolis, A.A., Manolis, A.S.: Cardiovascular safety of psychiatric agents: a cautionary tale. Angiology **70**(2), 103–129 (2019)
25. Li, K.J., Gurrera, R.J., Delisi, L.E.: Potentially fatal outcomes associated with clozapine. Schizophr. Res.. Res. **199**, 386–389 (2018)
26. Hoffman, K.B., et al.: The Weber effect and the United States Food and Drug Administration's Adverse Event Reporting System (FAERS): analysis of sixty-two drugs approved from 2006 to 2010. Drug Saf. **37**, 283–294 (2014)
27. Pu, G., et al.: Empirical analysis of the strengths and weaknesses of PEFT techniques for LLMs. arXiv preprint arXiv:2304.14999 (2023)
28. Kim, J., et al.: Memory-efficient fine-tuning of compressed large language models via sub-4-bit integer quantization. In: Advances in Neural Information Processing Systems 36 (2024)
29. Chai, Y., et al.: INT2.1: towards fine-tunable quantized large language models with error correction through low-rank adaptation. arXiv preprint arXiv:2306.08162 (2023)

30. Beijer, H.J.M., De Blaey, C.J.: Hospitalisations caused by Adverse Drug Reactions (ADR): a meta-analysis of observational studies. Pharm. World Sci. **24**(2), 46–54 (2002)
31. Pirmohamed, M., et al.: Adverse drug reactions as cause of admission to hospital: prospective analysis of 18 820 patients. BMJ **329**(7456), 15–19 (2004)
32. Onder, G., et al.: Adverse drug reactions as cause of hospital admissions: results from the Italian Group of Pharmacoepidemiology in the Elderly (GIFA). J. Am. Geriatr. Soc. **50**(12), 1962–1968 (2002)
33. Nelson, K.M., Talbert, R.L.: Drug-related hospital admissions. Pharmacother.: J. Hum. Pharmacol. Drug Ther. **16**, 701 (1996)
34. Bates, D.W.: Drugs and adverse drug reactions: how worried should we be?. JAMA **279**(15), 1216–1217 (1998)
35. Kodali, R.K., Upreti, Y.P., Boppana, L.: A quantization approach for the reduced size of large language models. In: 2024 16th International Conference on Knowledge and Smart Technology (KST). IEEE (2024)
36. Khin, N.A., et al.: Regulatory and scientific issues in studies to evaluate sexual dysfunction in antidepressant drug trials. J. Clin. Psychiatry **76**(8), 17222 (2015)
37. Sunkureddi, P., et al.: Using self-reported patient experiences to understand patient burden: learnings from digital patient communities in ankylosing spondylitis. Adv. Ther. **35**, 424–437 (2018)
38. Clark, E.M.: Applications in Sentiment Analysis and Machine Learning for Identifying Public Health Variables Across Social Media (2019)

ILP Models for String Partition Considering Intergenic Regions and Indels

Felipe Romeiro, Gabriel Siqueira(✉), Alexsandro Oliveira Alexandrino, and Zanoni Dias

Instituto de Computação, Universidade Estadual de Campinas (Unicamp), Campinas, Brazil
felipe.romeiro@students.ic.unicamp.br,
{gabriel.siqueira,alexsandro,zanoni}@ic.unicamp.br

Abstract. String partition problems have an important application in the development of algorithms for genome rearrangement distance problems. However, so far, the Integer Linear Programming approach has only been used for the simplest variation of string partition problems, where the representation of the genomes does not take the orientation of genes or intergenic regions into account, and the genomes must have the same set of genes. In this work, we propose Integer Linear Programming models for string partition problems incorporating information about gene orientation and intergenic regions, and allowing genomes with distinct gene sets. We also test the proposed models in simulated genomes.

Keywords: ILP · string partition · genome rearrangement

1 Introduction

Problems of genome rearrangement distance and string partitioning are studied in Computational Biology to estimate the evolutionary distance between two individuals. In the genome rearrangement distance problem, two genomes and a set of allowed operations are given, and the goal is to find the shortest sequence of operations that transforms the first genome into the second. Several genome rearrangement distance algorithms are based on the concept of string partitioning and motivated the variants of the problem studied in this work.

Strings may be used to represent genomes, with each gene represented by a character. Additionally, we can add other biologically relevant information to this representation, such as gene orientation and the amount of nucleotides between genes [13]. In this work, we use the term genomes to refer to the representation with additional information about intergenic regions. In the string partitioning problem, two strings or genomes are given, and the objective is to find the smallest set of substrings that can be concatenated to result in the input strings. Its most basic form is the Minimum Common Balanced String Partition (MCBSP), where one string is an anagram of the other. Without this restriction, we get the Minimum Common String Partition (MCSP). In some cases, it is useful to

T. Aldwairi et al. (Eds.): BICOB 2025, CCIS 2535, pp. 55–67, 2025.
https://doi.org/10.1007/978-3-031-94039-2_5

record the orientation of genes, which is usually done by marking a sign for each character, resulting in the Signed Minimum Common Balanced String Partition (SMCBSP) and the Signed Minimum Common String Partition (SMCSP).

In recent works, rearrangement distance problems have considered the amounts of nucleotides between genes, which we refer to as the sizes of intergenic regions. Initially, these works dealt with genomes with the same gene content [8–10,14,16], but more recent approaches are capable of dealing with genomes with unequal gene content [1–4,17]. Using this information, we have the Signed Minimum Common Intergenic String Partition (SMCISP). There is also a representation where the intergenic regions have a tolerance range [7,18], resulting in the Signed Minimum Common Flexible Intergenic String Partition (SMCFISP).

All these variants of partition problems are NP-hard [12,17,18], which motivates the exploration of Integer Linear Programming (ILP) models. Solutions for these partition problems are particularly interesting as components of algorithms for genome rearrangement distance problems considering rearrangement events such as reversals, which invert a segment of the genetic material; transpositions, which exchange two consecutive segments of the genetic material; or indels, which delete or insert segments of genetic material. The main benefit of developing algorithms with those solutions for partition problems is that we have the following theoretical guarantees:

- Solutions for SMCBSP are used in algorithms with approximation factor of 2 for genome rearrangement distance considering reversals [11].
- Solutions for MCBSP are used in algorithms with approximation factor of 2 for genome rearrangement distance considering transpositions [15].
- Solutions for SMICSP are used in algorithms with approximation factor of 4 for genome rearrangement distance considering reversals and indels, and approximation factor of 6 for genome rearrangement distance considering reversals, transpositions, and indels [17].
- Solutions for SMFICSP are used in algorithms with approximation factor of 4 for genome rearrangement considering reversals [18].

In the next section, we introduce some definitions and formally state the problems. In Sect. 3, we present the ILP formulations. In Sect. 4, we describe some experimental tests, and we conclude the paper in Sect. 5.

2 Definitions

A string S is a sequence of characters of size $|S|$, where the *i-th* character is denoted by S_i. In some cases, it may be interesting to identify the orientation of the characters, which is done by associating a sign ($+$ or $-$) to each of them. In this case, the string is referred to as *signed*, and *unsigned* otherwise.

The set of distinct characters in S is called the *alphabet* of S and is represented by Σ. To avoid ambiguity, the elements of Σ are referred to as *labels*, while the elements of S are called characters. In the case of signed strings, characters include the signs, and the labels do not, meaning that characters $+\alpha$ and $-\alpha$ refer to the label α.

Given two strings $S1$ and $S2$, if a label $\alpha \in \Sigma$ appears more times in $S1$ (respectively $S2$), it is said to be *abundant in* $S1$ (respectively *abundant in* $S2$). Otherwise, it is referred to as *rare*.

Definition 1. *Given two unsigned strings* $S1$ *and* $S2$*, a direct partition* $(\mathbb{S}1, \mathbb{S}2, \sigma)$ *of* $S1$ *and* $S2$ *consists of two sequences* $\mathbb{S}1$ *and* $\mathbb{S}2$ *of strings and a bijection* σ *between the elements of a subsequence* $\mathbb{S}1^\sigma$ *of* $\mathbb{S}1$ *and the elements of a subsequence* $\mathbb{S}2^\sigma$ *of* $\mathbb{S}2$ *such that the concatenation of the strings in* $\mathbb{S}1$ *is equal to* $S1$*, the concatenation of the strings in* $\mathbb{S}2$ *is equal to* $S2$*,* $\sigma(\mathbb{S}1_i) = \mathbb{S}2_j$ *for* $\mathbb{S}1_i \in \mathbb{S}1$ *and* $\mathbb{S}2_j \in \mathbb{S}2$*, only if* $\mathbb{S}1_i = \mathbb{S}2_j$*, and* $X \cap Y = \emptyset$*, considering that* X *is the set of characters in* $\mathbb{S}1$ *that are not in* $\mathbb{S}1^\sigma$ *and* Y *is the set of characters in* $\mathbb{S}2$ *that are not in* $\mathbb{S}2^\sigma$*.*

Example 1. A direct partition between two strings $S1$ and $S2$ without signs.

$$
\begin{aligned}
&S1 = (\ E\ \ B\ \ A\ \ C\ \ D\ \ E\ \ D\ )\\
&S2 = (\ B\ \ D\ \ C\ \ D\ \ E\ \ B\ \ E\ \ D\ \ D\ )\\
&\mathbb{S}1 = \langle (E\ \ B), (A), (C\ \ D), (E\ \ D) \rangle\\
&\mathbb{S}2 = \langle (B\ \ D), (C\ \ D), (E\ \ B), (E\ \ D), (D) \rangle\\
&\sigma(\mathbb{S}1_1) = \mathbb{S}2_3,\ \sigma(\mathbb{S}1_3) = \mathbb{S}2_2,\ \sigma(\mathbb{S}1_4) = \mathbb{S}2_4\\
&X = \{A\},\ Y = \{B\ \ D\}
\end{aligned}
$$

MINIMUM COMMON STRING PARTITION (*MCSP*)

Input: Two unsigned strings $S1$ and $S2$.
Goal: Find a direct partition $(\mathbb{S}1, \mathbb{S}2, \sigma)$ between $S1$ and $S2$ such that $|\mathbb{S}1| + |\mathbb{S}2 \setminus \mathbb{S}2^\sigma|$ is minimized.

Given a signed string S, a string $Q = inv(S)$ is called the *inverse* of S if it consists of the characters of S in reverse order and with the signs inverted, that is, $Q_i = -S_{|S|-i+1}, \forall 1 \leq i \leq |S|$.

Definition 2. *Given two signed strings* $S1$ *and* $S2$*, a signed partition* $(\mathbb{S}1, \mathbb{S}2, \sigma)$ *of* $S1$ *and* $S2$ *consists of two sequences* $\mathbb{S}1$ *and* $\mathbb{S}2$ *of strings and a bijection* σ *between the elements of a subsequence* $\mathbb{S}1^\sigma$ *of* $\mathbb{S}1$ *and the elements of a subsequence* $\mathbb{S}2^\sigma$ *of* $\mathbb{S}2$ *such that the concatenation of the strings in* $\mathbb{S}1$ *is equal to* $S1$*, the concatenation of the strings in* $\mathbb{S}2$ *is equal to* $S2$*,* $\sigma(\mathbb{S}1_i) = \mathbb{S}2_j$ *for* $\mathbb{S}1_i \in \mathbb{S}1$ *and* $\mathbb{S}2_j \in \mathbb{S}2$*, only if* $\mathbb{S}1_i = \mathbb{S}2_j$ *or* $S1_i = inv(S2_j)$*, and* $X \cap Y = \emptyset$*, considering that* X *is the set of characters in* $\mathbb{S}1$ *that are not in* $\mathbb{S}1^\sigma$ *and* Y *is the set of characters in* $\mathbb{S}2$ *that are not in* $\mathbb{S}2^\sigma$*.*

Example 2. A signed partition between two signed strings $S1$ and $S2$.

$$S1 = (\quad -A \quad -E \quad +B \quad -A \quad -C \quad +D \quad +E \quad -D \quad)$$
$$S2 = (\quad -D \quad +C \quad +A \quad -E \quad +B \quad +D \quad -E \quad)$$
$$\mathbb{S}1 = \langle(-A), (-E \quad +B), (-A \quad -C \quad +D), (+E \quad -D)\rangle$$
$$\mathbb{S}2 = \langle(-D \quad +C \quad +A), (-E \quad +B), (+D \quad -E)\rangle$$
$$\sigma(\mathbb{S}1_2) = \mathbb{S}2_2,\ \sigma(\mathbb{S}1_3) = \mathbb{S}2_1,\ \sigma(\mathbb{S}1_4) = \mathbb{S}2_3$$
$$X = \{A\},\ Y = \emptyset$$

SIGNED MINIMUM COMMON STRING PARTITION (*SMCSP*)
Input: Two signed strings $S1$ and $S2$.
Goal: Find a signed partition $(\mathbb{S}1, \mathbb{S}2, \sigma)$ of $S1$ and $S2$ such that $\lvert\mathbb{S}1\rvert + \lvert\mathbb{S}2 \setminus \mathbb{S}2^\sigma\rvert$ is minimized.

In the next problems, a genome $G = (S, \breve{S})$ is composed of a string S and a sequence of intergenic regions $\breve{S}$ of length $|S| - 1$. The genome is said to be rigid if $\breve{S}$ is a sequence of natural numbers. Given two genomes $G1 = (S1, \breve{S1})$ and $G2 = (S2, \breve{S2})$, $G2 = inv(G1)$ is said to be the inverse of $G1$ if $S2 = inv(S1)$ and $\breve{S2}$ is the sequence $\breve{S1}$ in reverse order, that is, $\breve{S2}_i = \breve{S1}_{|S1|-i}$, $\forall 1 \leq i \leq |S1| - 1$.

A *break* of a genome $G1 = (S1, \breve{S1})$ is an operation that splits $G1$ into two genomes $G1' = (S1', \breve{S1}')$ and $G1'' = (S1'', \breve{S1}'')$ at an intergenic region $\breve{S1}_i$ such that: $|S1'| = i$; $S1'_j = S1_j, \forall 1 \leq j \leq i$; $|S1''| = |S1| - i$; $S1''_{j-i} = S1_j, \forall i < j \leq |S1|$; $|\breve{S1}'| = i$; $\breve{S1}'_j = \breve{S1}_j, \forall 1 \leq j < i$; $|\breve{S1}''| = |\breve{S1}| - i$; $\breve{S1}''_{j-i} = \breve{S1}_j, \forall i < j \leq |S1| - 1$.

Definition 3. *Given two rigid genomes* $G1 = (S1, \breve{S1})$ *and* $G2 = (S2, \breve{S2})$*, a signed partition* $(\mathbb{S}1, \mathbb{S}2, \sigma)$ *is composed of two sequences of genomes* $\mathbb{S}1$ *and* $\mathbb{S}2$ *and a bijection* σ *between the elements of a subsequence* $\mathbb{S}1^\sigma$ *of* $\mathbb{S}1$ *and of a subsequence* $\mathbb{S}2^\sigma$ *of* $\mathbb{S}2$ *such that the genome* $G1$ *can be broken into the genomes of* $\mathbb{S}1$*, the genome* $G2$ *can be broken into the genomes of* $\mathbb{S}2$*,* $\sigma(\mathbb{S}1_i) = \mathbb{S}2_j$ *for* $\mathbb{S}1_i \in \mathbb{S}1$ *and* $\mathbb{S}2_j \in \mathbb{S}2$ *only if* $\mathbb{S}1_i = \mathbb{S}2_j$ *or* $\mathbb{S}1_i = inv(\mathbb{S}2_j)$*, and* $X \cap Y = \emptyset$*, considering that* X *is the set of labels of the characters in* $\mathbb{S}1$ *that are not in* $\mathbb{S}1^\sigma$ *and* Y *is the set of labels of the characters in* $\mathbb{S}2$ *that are not in* $\mathbb{S}2^\sigma$*.*

Example 3. A signed partition between two rigid genomes $(S1, \breve{S1})$ and $(S2, \breve{S2})$.

$$S1 = (\quad +A \quad +E \quad -B \quad +D \quad) \qquad \breve{S1} = (2 \quad 3 \quad 1)$$
$$S2 = (\quad +E \quad -B \quad +D \quad -A \quad) \qquad \breve{S2} = (3 \quad 2 \quad 1)$$
$$\mathbb{S}1 = \langle((+A), ()), ((+E \quad -B), (3)), ((+D), ())\rangle$$
$$\mathbb{S}2 = \langle((+E \quad -B), (3)), ((+D), ()), ((-A), ())\rangle$$
$$\sigma(\mathbb{S}1_1) = \mathbb{S}2_3,\ \sigma(\mathbb{S}1_2) = \mathbb{S}2_1,\ \sigma(\mathbb{S}1_3) = \mathbb{S}2_2$$
$$X = \emptyset,\ Y = \emptyset$$

Signed Minimum Common Intergenic String Partition (*SMCISP*)
Input: Two rigid genomes $G1 = (S1, \breve{S}1)$ and $G2 = (S2, \breve{S}2)$.   **Goal:** Find a signed partition $(\mathbb{S}1, \mathbb{S}2, \sigma)$ of $G1$ and $G2$ such that $\lvert\mathbb{S}1\rvert + \lvert\mathbb{S}2 \setminus \mathbb{S}2^\sigma\rvert$ is minimized.

A flexible genome $G = (S, \breve{S})$ is composed of a string S and a sequence of pairs of natural numbers $\breve{S}$ of size $|S| - 1$ such that $\breve{S}_i = (\breve{S}_i^{min}, \breve{S}_i^{max})$ and $\breve{S}_i^{min} \leq \breve{S}_i^{max}$. Given a rigid genome $G1 = (S1, \breve{S}1)$ and a flexible genome $G2 = (S2, \breve{S}2)$, it is said that $G1$ and $G2$ are compatible if $S1 = S2$ and $\breve{S}2_i^{min} \leq \breve{S}1_i \leq \breve{S}2_i^{max}$, $\forall 1 \leq i \leq |S1| - 1$.

Definition 4. *Given a rigid genome $G1 = (S1, \breve{S}1)$ and a flexible genome $G2 = (S2, \breve{S}2)$, a partition with signs $(\mathbb{S}1, \mathbb{S}2, \sigma)$ is composed of a sequence of rigid genomes $\mathbb{S}1$, a sequence of flexible genomes $\mathbb{S}2$ and a bijection σ between the elements of a subsequence $\mathbb{S}1^\sigma$ of $\mathbb{S}1$ and the elements of a subsequence $\mathbb{S}2^\sigma$ of $\mathbb{S}2$ such that the genome $G1$ can be broken into the genomes of $\mathbb{S}1$, the genome $G2$ can be broken into the genomes of $\mathbb{S}2$, $\sigma(\mathbb{S}1_i) = \mathbb{S}2_j$ for $\mathbb{S}1_i \in \mathbb{S}1$ and $\mathbb{S}2_j \in \mathbb{S}2$ only if $\mathbb{S}1_i$ is compatible with $\mathbb{S}2_j$ or with $inv(\mathbb{S}2_j)$, and $X \cap Y = \emptyset$, considering that X is the set of labels of the characters in $\mathbb{S}1$ that are not in $\mathbb{S}1^\sigma$ and Y is the set of labels of the characters in $\mathbb{S}2$ that are not in $\mathbb{S}2^\sigma$.*

Example 4. A partition with signs between a rigid genome $(S1, \breve{S}1)$ and a flexible genome $(S2, \breve{S}2)$.

$$S1 = (\ +A \quad -E \quad +B \quad -A \quad -C \quad +D \quad +E \quad -D\ )$$
$$S2 = (\ -D \quad +C \quad +A \quad -E \quad +B \quad +D \quad -E\ )$$
$$\breve{S}1 = (\ 63 \quad 44 \quad 23 \quad 52 \quad 32 \quad 38 \quad 4\ )$$
$$\breve{S}2 = (\ (25,47) \quad (0,54) \quad (63,64) \quad (42,45) \quad (34,68) \quad (0,10)\ )$$
$$\mathbb{S}1 = \langle((+A),()),((-E \quad +B),(44)),((-A \quad -C \quad +D),(52 \quad 32)), (+E \quad -D,(4))\rangle$$
$$\mathbb{S}2 = \langle((-D \quad +C \quad +A,((25,47) \quad (0,54))),((-E \quad +B),((42,45)), ((+D \quad -E),((0,10)))\rangle$$
$$\sigma(\mathbb{S}1_2) = \mathbb{S}2_2,\ \sigma(\mathbb{S}1_3) = \mathbb{S}2_1,\ \sigma(\mathbb{S}1_4) = \mathbb{S}2_3$$
$$X = \{A\},\ Y = \emptyset$$

Signed Minimum Common Flexible Intergenic String Partition (*SMCFISP*)
Input: A rigid genome $G1$ and a flexible genome $G2$.   **Goal:** Find a partition with signs $(\mathbb{S}1, \mathbb{S}2, \sigma)$ of $G1$ and $G2$ such that $\lvert\mathbb{S}1\rvert + \lvert\mathbb{S}2 \setminus \mathbb{S}2^\sigma\rvert$ is minimal.

3 Integer Linear Programming Formulations

This section presents integer linear formulations for the proposed partition problems by adapting to formulations proposed in the literature for partition problems without signs or intergenic regions. The first formulation is based on common blocks [5], and will be denoted by **CB**. The second formulation is based on common strings [6], and will be denoted by **CS**.

3.1 Minimum Common String Partition

Given two input strings $S1$ and $S2$, each common block b_i is represented by a triple $(t_i, k1_i, k2_i)$ composed of a string t_i and two indices $1 \leq k1_i, k2_i \leq |S1|$, indicating that the substring t_i starts at position $k1_i$ in $S1$ and at position $k2_i$ in $S2$. Let B be the set of all possible common blocks between $S1$ and $S2$. Denote a possible exclusive block $e1_i$ of $S1$ by the tuple $(r1_i, \ell1_i)$, where $r1_i$ is a substring of characters with abundant labels in $S1$ and $\ell1_i$ is the position where this substring starts in $S1$. Denote an exclusive block $e2_i = (r2_i, \ell2_i)$ of $S2$ similarly. Let $E1$ and $E2$ be the sets of all possible exclusive blocks of $S1$ and $S2$, respectively. The following theorem ensure us that we can use this sets of exclusive blocks to represent the exclusive parts of any partition.

Theorem 1. *Given a partition with signs* $(\mathbb{S}1, \mathbb{S}2, \sigma)$*, the exclusive parts are formed only by characters with abundant labels.*

Proof. By contradiction and without loss of generality, assume there exists an exclusive part $\mathbb{S}1_i$ that contains a character with a rare label α. By definition of rare label, α appears at least $occ(\alpha, \mathbb{S}1)$ times in $\mathbb{S}2$. However, the common (non-exclusive) parts of $\mathbb{S}1$ and $\mathbb{S}2$ contain at most $occ(\alpha, \mathbb{S}1) - 1$ instances of α. Thus, the exclusive parts of $\mathbb{S}2$ also have at least one instance of α. This implies a contradiction with the definition of partition which requires that characters of exclusive parts of $\mathbb{S}1$ do not appear in the exclusive parts of $\mathbb{S}2$. Therefore, the exclusive parts are formed only by characters with abundant labels.

The **CB** model uses, for each block $b_i \in B$, a binary variable x_i that indicates whether this block is part of the solution S. Additionally, we have binary variables $y1_{r1_i,\ell1_i}$, for each $e1_i \in E1$, and $y2_{r2_i,\ell2_i}$, for each $e2_i \in E2$, to represent the exclusive blocks. With these variables, the model can be written as follows:

$$\text{min.} \quad \sum_{b_i \in B} x_i + \sum_{e1_i \in E1} y1_i + \sum_{e2_i \in E2} y2_i \tag{1}$$

$$\text{s.t.} \quad \sum_{b_i \in B, k1_i \leq j < k1_i + |t_i|} x_i + \sum_{e1_i \in E1, \ell1_i \leq j < \ell1_i + |r1_i|} y1_i = 1, \forall 1 \leq j \leq |S1| \tag{2}$$

$$\sum_{b_i \in B, k2_i \leq j < k2_i + |t_i|} x_i + \sum_{e2_i \in E2, \ell2_i \leq j < \ell2_i + |r2_i|} y2_i = 1, \forall 1 \leq j \leq |S2| \tag{3}$$

$$x_i \in \{0,1\}, \forall b_i \in B, \quad y1_i \in \{0,1\}, \forall e1_i \in E1, \quad y2_i \in \{0,1\}, \forall e2_i \in E2$$

Objective function (1) ensures that the found solution is minimal. Note that this function corresponds to the objective of the problem, because $\sum_{b_i \in B} x_i + \sum_{e1_i \in E1} y1_i$ contains all parts of $S1$ ($\mathbb{S}1$), and $\sum_{e2_i \in E2} y2_i$ contains the exclusive parts of $S2$ ($\mathbb{S}2 \setminus \mathbb{S}2^{\sigma}$). Constraint (2) ensures that each character of $S1$ is represented by one and only one block of $S1$. Constraint (3) is analogous for $S2$. Note that if a label of a character is abundant in one of the strings, it is necessarily rare in the other. This implies that a label can only appear in the exclusive blocks of one of the strings, and thus, the set of blocks found by this formulation satisfies the partition definition.

A major problem with this formulation is that repeated substrings generate many blocks, increasing the size of the instance. The **CS** attempts to reduce this problem. Given two input strings $S1$ and $S2$, let T be the set of all substrings common to $S1$ and $S2$. For each $t \in T$, let $Q1_t$ (respectively $Q2_t$) be the set of positions where t starts in $S1$ (respectively $S2$). For each combination of $t \in T$ and $k \in Q1_t$, this model uses a binary variable $x1_{t,k}$ to indicate that the substring t starting at k in string $S1$ is part of the solution. The variables $x2_{t,k}$ are analogous for $S2$. Thus, the model can be written as follows:

$$\text{min.} \quad \sum_{t \in T} \sum_{k \in Q1_t} x1_{t,k} + \sum_{e1_i \in E1} y1_i + \sum_{e2_i \in E2} y2_i \tag{4}$$

$$\text{s.t.} \quad \sum_{t \in T} \sum_{k \in Q1_t, k \leq j < k+|t|} x1_{t,k} + \sum_{e1_i \in E1, \ell1_i \leq j < \ell1_i + |r1_i|} y1_j = 1, \forall 1 \leq j \leq |S1| \tag{5}$$

$$\sum_{t \in T} \sum_{k \in Q2_t, k \leq j < k+|t|} x2_{t,k} + \sum_{e2_i \in E2, \ell2_i \leq j < \ell2_i + |r2_i|} y2_i = 1, \forall 1 \leq j \leq |S2| \tag{6}$$

$$\sum_{k \in Q1_t} x1_{t,k} = \sum_{k \in Q2_t} x2_{t,k}, \forall t \in T \tag{7}$$

$$x1_{t,k} \in \{0,1\}, \forall t \in T, k \in Q1_t, \quad x2_{t,k} \in \{0,1\}, \forall t \in T, k \in Q2_t$$

$$y1_i \in \{0,1\}, \forall e1_i \in E1, \quad y2_i \in \{0,1\}, \forall e2_i \in E2$$

Again, the objective function (4) corresponds to the one in MCSP, and constraints (5) and (6) ensure that each character is part of exactly one of the selected blocks. Constraint (7), ensures that the common blocks are chosen in equal numbers in each of the strings, and therefore it is possible to pair them and build a partition from the selected strings.

3.2 Signed Minimum Common String Partition

The formulation **CB** remains as in the previous case, but using the concept of *common blocks with signs*. Given two strings with signs $S1$ and $S2$, a common block with signs b_i is represented by the triple $(t_i, k1_i, k2_i)$ composed of a string t_i and two indices $1 \leq k1_i \leq |S1|$ and $1 \leq k2_i \leq |S2|$, indicating that $S1$ has either the substring t_i or $inv(t_i)$ at position $k1_i$ (similarly for $S2$ and $k2_i$). Other concepts remain unchanged.

The formulation **CS** undergoes a similar modification. Given two strings $S1$ and $S2$, let T be the set of all strings t such that t or $inv(t)$ is a substring of both $S1$ and $S2$. For each $t \in T$, let $Q1_t$ (respectively $Q2_t$) be the set of positions where t starts in $S1$ (respectively $S2$). The rest of the formulation remains the same as the last case.

3.3 Signed Minimum Intergenic Common String Partition

For the **CB** model, it will be necessary to introduce the concept of common blocks of intergenic regions. Given two rigid genomes $G1 = (S1, \breve{S}1)$ and $G2 = (S2, \breve{S}2)$, a common block is represented by the triple $(t_i, k1_i, k2_i)$ consisting of a genome t_i and two indices $1 \leq k1_i \leq |S1|$ and $1 \leq k2_i \leq |S2|$, indicating that $S1$ has the genome t_i or $inv(t_i)$ at position $k1_i$ (similarly for $S2$ and $k2_i$). The rest of the formulation remains as before.

For the **CS** formulation, greater changes are needed. Given a rigid genome $G1 = (S1, \breve{S}1)$ and a flexible genome $G2 = (S2, \breve{S}2)$, let T be the set of all genomes contained in $G1$ that are compatible with some genome contained in $G2$. For $t \in T$, let $Q1_t$ be the set of positions where t occurs in $G1$, and $Q2_t$ be the set of positions where a genome compatible with t or $inv(t)$ appears in $G2$. The rest of the formulation remains the same as before.

3.4 Signed Flexible Intergenic Common String Partition

For the **CB** formulation, given a rigid genome $G1 = (S1, \breve{S}1)$ and a flexible genome $G2 = (S2, \breve{S}2)$, a common block is represented by the triple $(t_i, k1_i, k2_i)$ indicating that the rigid genome t_i appears in $G1$ at position $k1_i$ and is compatible with the flexible genome starting at position $k2_i$ in $G2$ or its inverse. The rest of the formulation remains unchanged.

The **CS** formulation remains as before, with the inclusion of the following changes. Given a rigid genome $G1 = (S1, \breve{S}1)$ and a flexible genome $G2 = (S2, \breve{S}2)$, T is the set of all genomes in $G1$ that are compatible with some genome in $G2$. For $t \in T$, $Q1_t$ is the set of positions where t occurs in $G1$, and $Q2_t$ is the set of positions where a genome compatible with t or $inv(t)$ appears in $G2$.

4 Experiments

The models described earlier and their adaptations were developed using Python 3.12.2 and solved using Gurobi 11. The following tests were conducted on a computer equipped with an "Intel(R) Xeon(R) CPU E5-2470 v2" with 40 cores at 2.40GHz, 32GB of RAM, and 9GB of swap space. Each test was run in a single thread, where model construction had no time limits, and model execution was capped at 1 h.

To create the instances, two parameters were iterated: the size of the alphabet $|\Sigma|$ from $\{4, 20, 36\}$ and the fraction of conservative operations c from $\{0.6, 0.8, 1\}$. For each parameter combination, 10 tests were created. In each instance, a rigid genome with l labels and a length of 1000 was generated, and 1000 operations were executed: $1000c$ conservative operations (all reversals) and $1000(1-c)$ indels (deletions followed by insertions) were applied to obtain the second genome. The original rigid genome was created by uniformly randomly selecting genes from an alphabet Σ and intergenic regions with values in the range $[0, 100]$. Reversal operations randomly and uniformly select two positions in the genome and invert the segment of genes and intergenic regions between these positions, reversing the signs of affected genes. Each deletion removes a randomly chosen gene and adjacent nucleotides, and each insertion inserts a gene with a label not present in the genome at a randomly chosen position, along with a random number of nucleotides around the gene. At most 100 nucleotides are inserted in each region.

These instances reflect the SMCISP. For SMCSP, the intergenic regions were discarded. For SMCFISP, two tolerances $v \in \{0.25, 0.5\}$ were defined. For each instance $(G1, G2 = (S2, \breve{S}2))$, two flexible genomes $G2' = (S2', \breve{S}2')$ were constructed such that for each $G2_i \in G2$, $\breve{S}2'^{min}_i = (1-v)\breve{S}2_i$ and $\breve{S}2'^{max}_i = (1+v)\breve{S}2_i$, with $S2' = S2$.

The Table 1 list the results of tests for the **CB** model and the Table 2 list the results of tests for the for the **CS** model. It can be observed that tests with more characters tend to run faster and achieve a smaller gap (distance between the lower and upper bound found by the solver if the exact solution was not yet found), which is expected because there are fewer common substrings between inputs, resulting in smaller models that are therefore easier to optimize. Furthermore, it is noted that in both models, tests with 80% and 100% conservative operations tend to run faster than tests with 60%, suggesting that exclusive blocks are computationally more expensive than common blocks. In the case of model CB, this effect is more noticeable in tests with 36 labels, while in model CS, this effect can be observed in almost all tests, and in both cases, execution times with 100% conservative operations are nearly negligible. This suggests that improvements in the treatment of exclusive blocks would be advantageous for the studied models. One possible improvement would be to consider only exclusive blocks that are adjacent to either a common block or to one end of the string.

The results of model **CB** were inconclusive regarding the impact of flexible intergenic region tolerance on model execution: there are cases where tests with lower tolerance performed better and others where tests with higher tolerance were better. Nevertheless, in the instances produced with 80% conservative operations and alphabet of size 4, where the gap is more noticeable, we can see that larger values of tolerance lead to large values of gap. Besides, it can be observed that both in the case of 60% and 80% conservative operations, SMCISP tests with 4 characters ran faster than the others, which is curious because, on the one hand, SMCISP tests are expected to run faster than others due to having fewer common blocks, but SMCISP tests with more characters ran for longer.

Table 1. Results of tests for the **CB** model. Tolerance 0 indicates SMCISP, tolerances 25 and 50 indicates SMCFISP, and tolerance ∞ indicates SMCSP.

$\|\Sigma\|$	Conservative Operations (%)	Tolerance (%)	Model Time (s)		Gap (%)		Total Time (s)	
			Mean	SD	Mean	SD	Mean	SD
4	60	0	2103	284	0.0	0.0	3266	440
4	60	25	3540	204	1.6	0.8	4687	619
4	60	50	3381	480	0.6	0.4	4466	482
4	60	∞	3415	615	0.8	0.5	4503	619
20	60	0	3214	715	0.4	0.5	4230	718
20	60	25	2490	751	0.0	0.0	3518	735
20	60	50	2384	751	0.0	0.0	3397	750
20	60	∞	2824	892	0.2	0.4	3841	902
36	60	0	3603	1	0.5	0.2	4603	9
36	60	25	3471	416	0.3	0.2	4473	415
36	60	50	3472	412	0.3	0.3	4470	412
36	60	∞	3065	783	0.2	0.3	4071	785
4	80	0	1866	255	0.0	0.0	2982	257
4	80	25	3603	2	3.7	6.6	4719	26
4	80	50	3603	2	6.1	7.6	4746	94
4	80	∞	3607	14	12.6	12.2	4734	22
20	80	0	3253	396	0.1	0.5	4264	398
20	80	25	2618	707	0.1	0.2	3626	706
20	80	50	2169	445	0.0	0.0	3178	436
20	80	∞	3542	190	9.4	19.3	4549	189
36	80	0	3247	1125	0.3	0.2	4160	1437
36	80	25	2929	1129	0.2	0.2	3839	1408
36	80	50	2333	1127	0.1	0.1	3247	1366
36	80	∞	2431	1062	0.0	0.0	3349	1323
4	100	0	2	1	0.0	0.0	138	1
4	100	25	3600	0	1.1	0.3	3740	1
4	100	50	3600	0	2.1	0.5	3744	1
4	100	∞	3600	0	4.0	0.8	3751	2
20	100	0	3163	378	0.1	0.3	4267	380
20	100	25	2884	379	0.1	0.3	3892	378
20	100	50	2578	379	0.1	0.1	3584	379
20	100	∞	2533	378	0.1	0.1	3541	379
36	100	0	2510	377	0.1	0.1	3541	378
36	100	25	2274	377	0.0	0.0	3347	378
36	100	50	2294	377	0.0	0.0	3367	378
36	100	∞	2189	378	0.0	0.0	3269	379

Table 2. Results of tests for the **CS** model. Tolerance 0 indicates SMCISP, tolerances 25 and 50 indicates SMCFISP, and tolerance ∞ indicates SMCSP.

$\|\Sigma\|$	Conservative Operations (%)	Tolerance (%)	Model Time (s)		Gap (%)		Total Time (s)	
			Mean	SD	Mean	SD	Mean	SD
4	60	0	2	1	0.0	0.0	132	1
4	60	25	3516	14	3.6	0.1	3650	14
4	60	50	3524	11	6.3	0.2	3658	11
4	60	∞	3535	24	12.7	0.2	3669	24
20	60	0	3221	381	0.2	0.2	4295	382
20	60	25	2820	372	0.2	0.2	3866	372
20	60	50	2351	373	0.0	0.0	3380	372
20	60	∞	2338	373	0.0	0.0	3366	372
36	60	0	2338	373	0.0	0.0	3366	372
36	60	25	2197	373	0.0	0.0	3240	372
36	60	50	2136	372	0.0	0.0	3175	371
36	60	∞	2161	372	0.0	0.0	3200	372
4	80	0	3478	12	0.0	0.0	3612	13
4	80	25	3600	0	4.5	0.0	3714	0
4	80	50	3600	0	7.2	0.0	3714	1
4	80	∞	3600	0	12.6	0.0	3713	2
20	80	0	3213	376	0.2	0.3	4288	378
20	80	25	2835	378	0.1	0.2	3863	378
20	80	50	2518	377	0.1	0.1	3550	377
20	80	∞	2552	378	0.1	0.1	3584	378
36	80	0	2552	378	0.1	0.1	3584	378
36	80	25	2414	378	0.0	0.0	3452	378
36	80	50	2326	378	0.0	0.0	3365	378
36	80	∞	2347	378	0.0	0.0	3386	378
4	100	0	3	2	0.0	0.0	132	1
4	100	25	3600	0	0.9	0.2	3740	1
4	100	50	3600	0	2.3	0.3	3744	1
4	100	∞	3600	0	3.9	0.2	3749	1
20	100	0	3137	372	0.0	0.1	4203	373
20	100	25	2801	371	0.0	0.0	3829	372
20	100	50	2497	372	0.0	0.0	3526	372
20	100	∞	2488	372	0.0	0.0	3516	372
36	100	0	2488	372	0.0	0.0	3516	372
36	100	25	2362	371	0.0	0.0	3390	371
36	100	50	2296	371	0.0	0.0	3324	371
36	100	∞	2315	371	0.0	0.0	3342	371

In model **CS**, the impact of flexible intergenic region tolerance is more noticeable, which is expected since higher tolerance allows for the construction of more common blocks. It is also interesting to note that this model is better than the previous one in all scenarios, running faster and achieving smaller gaps.

We can also observe that if we have only conservative operations, the model assembly time (difference between total time and model time) was about 1000 s, which, considering the execution time of 1 h, is reasonable. Moreover, in all scenarios, the standard deviation of the model execution time frequently approached the standard deviation of the total time, indicating a consistency that suggests that the modeling step is minimally affected by the instances and that most of the variation in time results from the model execution phase.

5 Conclusion

This work has demonstrated that the formulations CB and CS for MCBSP, introduced in previous works, can be adapted to solve common variants of the string partition problem. The models considering these adaptations solve the tested instances with quality responses, given the chosen execution time for the tests. In some cases, the instances were optimally solved within the time limit.

In future work, exploration of other models for the proposed problems could be pursued. It may also be interesting to study heuristics to accelerate the solution time of the developed models. Another possibility, given the impact of exclusive blocks on the execution times of the models, would be to attempt to reduce the quantity of these blocks.

Acknowledgement. This work was supported by the São Paulo Research Foundation, FAPESP (grant 2021/13824-8).

References

1. Alexandrino, A.O., Brito, K.L., Oliveira, A.R., Dias, U., Dias, Z.: Reversal and indel distance with intergenic region information. IEEE/ACM Trans. Comput. Biol. Bioinform. 1–13 (2022). https://doi.org/10.1109/TCBB.2022.3215615
2. Alexandrino, A.O., Oliveira, A.R., Dias, U., Dias, Z.: Incorporating intergenic regions into reversal and transposition distances with indels. J. Bioinform. Comput. Biol. **19**(06), 2140,011 (2021)
3. Alexandrino, A.O., Oliveira, A.R., Jean, G., Fertin, G., Dias, U., Dias, Z.: Transposition distance considering intergenic regions for unbalanced genomes. In: International Symposium on Bioinformatics Research and Applications (ISBRA), vol. 13760, pp. 100–113 (2022)
4. Alexandrino, A.O., Oliveira, A.R., Jean, G., Fertin, G., Dias, U., Dias, Z.: Reversal and transposition distance on unbalanced genomes using intergenic information. J. Comput. Biol. **30**(8), 861–876 (2023)
5. Blum, C., Lozano, J.A., Davidson, P.: Mathematical programming strategies for solving the minimum common string partition problem. Eur. J. Oper. Res. **242**(3), 769–777 (2015)

6. Blum, C., Raidl, G.R.: Computational performance evaluation of two integer linear programming models for the minimum common string partition problem. Optim. Lett. **10**(1), 189–205 (2015). https://doi.org/10.1007/s11590-015-0921-4
7. Brito, K.L., Alexandrino, A.O., Oliveira, A.R., Dias, U., Dias, Z.: Genome rearrangement distance with a flexible intergenic regions aspect. IEEE/ACM Trans. Comput. Biol. Bioinf. **20**(03), 1641–1653 (2023)
8. Brito, K.L., Jean, G., Fertin, G., Oliveira, A.R., Dias, U., Dias, Z.: Sorting by genome rearrangements on both gene order and intergenic sizes. J. Comput. Biol. **27**(2), 156–174 (2020)
9. Brito, K.L., Oliveira, A.R., Alexandrino, A.O., Dias, U., Dias, Z.: An improved approximation algorithm for the reversal and transposition distance considering gene order and intergenic sizes. Algorithms Mol. Biol. **16**(1), 1–21 (2021)
10. Bulteau, L., Fertin, G., Tannier, E.: Genome rearrangements with indels in intergenes restrict the scenario space. BMC Bioinform. **17**(14), 426 (2016)
11. Chen, X., et al.: Assignment of orthologous genes via genome rearrangement. IEEE/ACM Trans. Comput. Biol. Bioinf. **2**(4), 302–315 (2005)
12. Goldstein, A., Kolman, P., Zheng, J.: Minimum common string partition problem: hardness and approximations. In: Fleischer, R., Trippen, G. (eds.) ISAAC 2004. LNCS, vol. 3341, pp. 484–495. Springer, Heidelberg (2004). https://doi.org/10.1007/978-3-540-30551-4_43
13. Oliveira, A.R., Brito, K.L., Alexandrino, A.O., Siqueira, G., Dias, U., Dias, Z.: Rearrangement distance problems: an updated survey. ACM Comput. Surv. **56**(8) (2024)
14. Oliveira, A.R., Jean, G., Fertin, G., Brito, K.L., Dias, U., Dias, Z.: Sorting permutations by intergenic operations. IEEE/ACM Trans. Comput. Biol. Bioinf. **18**(6), 2080–2093 (2021)
15. Shapira, D., Storer, J.A.: Edit distance with move operations. J. Discret. Algorithms **5**(2), 380–392 (2007)
16. Siqueira, G., Alexandrino, A.O., Dias, Z.: Signed rearrangement distances considering repeated genes and intergenic regions. In: Proceedings of 14th International Conference on Bioinformatics and Computational Biology (BICoB 2022), vol. 83, pp. 31–42 (2022)
17. Siqueira, G., Alexandrino, A.O., Dias, Z.: Signed rearrangement distances considering repeated genes, intergenic regions, and indels. J. Comb. Optim. **46**(2), 16 (2023)
18. Siqueira, G., Alexandrino, A.O., Oliveira, A.R., Jean, G., Fertin, G., Dias, Z.: Approximating rearrangement distances with replicas and flexible intergenic regions. In: International Symposium on Bioinformatics Research and Applications (ISBRA), vol. 14248, pp. 241–254 (2023)

Beyond AlphaFold: Unraveling FGFR2-Related Diseases with Enhanced Computational Workflow

Yiyang Lian[1], Dale Bodian[2], and Amarda Shehu[1,3](✉)

[1] School of Systems Biology, George Mason University, Manassas, VA 20110, USA
{ylian,amarda}@gmu.edu
[2] Diamond Age Data Science, Boston, MA, USA
[3] Department of Computer Science, George Mason University, Fairfax, VA 22030, USA

Abstract. Methods like AlphaFold2, AlphaMissense, and AlphaFold3 are advancing our ability to link protein sequences to functions by providing accurate 3D models of protein structures, surpassing limitations of wet-lab methods. Two questions remain: How accurate are these methods for specific proteins, and what is the optimal workflow, especially considering the importance of structural dynamics, which current AlphaFold methods do not capture? This study focuses on Fibroblast Growth Factor Receptor 2 (FGFR2), a protein critical in cellular processes and associated with various congenital disorders and cancers. We evaluate different workflows for predicting the structural and functional impacts of FGFR2 mutations, finding that a sequence involving AlphaFold2 (structure prediction), followed by SCWRL4 (side-chain optimization) and NAMD2 (structure dynamics), achieves high accuracy and offers insights into both dynamic structural and functional changes. This approach advances our understanding of FGFR2-related diseases and more broadly provides a reliable framework for predicting the structural and functional impact of single-nucleotide variants.

Keywords: FGFR2 · Protein Mutant Modeling · AlphaFold2 · Molecular Dynamics · Structural Bioinformatics

1 Introduction

Accurate modeling of protein mutants, particularly for specific proteins such as FGFR2 (fibroblast growth factor receptor 2), is essential to understand their function and the molecular mechanisms underlying disease [1–3]. FGFR2 plays a critical role in human biology and mutations in this protein can cause significant structural and functional changes that are involved in various health conditions [4–8]. Therefore, it is crucial to develop robust computational methods to predict how sequence variations, such as single nucleotide variants (SNVs), impact protein structure and function.

Several cutting-edge approaches have been developed to predict the three-dimensional structures of proteins based on their sequence information [9–11]. In addition to classical homology-based or threading-based methods, recent advancements have

T. Aldwairi et al. (Eds.): BICOB 2025, CCIS 2535, pp. 68–84, 2025.
https://doi.org/10.1007/978-3-031-94039-2_6

introduced deep learning-based approaches such as AlphaFold2, which relies on multiple sequence alignments, and ESMFold, which, although it does not use sequence alignment, achieves a comparable level of predictive accuracy. Although these methods, including AlphaFold2 and AlphaMissense, can provide reasonable three-dimensional structure templates, they have shown limited sensitivity when applied to specific proteins such as FGFR2. Studies on receptor tyrosine kinases (RTKs) indicate that AlphaFold2 tends to over-stabilize compact structures and fails to capture activation-induced conformational shifts [12, 13]. Molecular dynamics simulations help overcome this limitation by explicitly modeling the flexibility of key functional regions, such as the activation loop and kinase hinge, providing a more comprehensive understanding of the mutational impact on FGFR2 function. This insensitivity becomes particularly evident when attempting to understand how sequence changes in the form of SNVs propagate through structural alterations and ultimately lead to functional changes or dysfunction. For instance, AlphaFold2 tends to maintain a bias towards an average compact structure, even when mutations are introduced, failing to capture subtle but critical structural variations.

Moreover, current structure prediction methods typically do not incorporate dynamic information, yet the relationship between sequence mutations and functional outcomes is often mediated by structural dynamics. This is especially important when focusing on specific proteins such as FGFR2, where precise predictions linking SNVs to functional changes are necessary. Neglecting the role of dynamics in these predictions can lead to incomplete or inaccurate models, hindering our understanding of the mutation-function relationship.

However, relying entirely on classical molecular mechanics methods, such as molecular dynamics (MD) simulations [14], to capture these dynamics is computationally infeasible for many practical applications. Therefore, the question this paper addresses is how to effectively combine state-of-the-art structure prediction methods with rapid incorporation of dynamic information to more accurately link sequence mutations to functional changes.

This question is particularly challenging in molecular biology and we explore it in the context of FGFR2 kinase, a protein central to both human biology and health. FGFR2, a member of the fibroblast growth factor receptor family, is involved in crucial biological processes such as cell growth, differentiation, and angiogenesis. Proper FGFR2 function is vital for normal development, and mutations in this protein are associated with various congenital disorders and cancers, including craniosynostosis syndromes and certain types of breast and gastric cancers. Understanding how mutations in FGFR2 lead to structural and functional changes is of paramount importance, not only to unravel the molecular mechanisms of these diseases, but also to identify potential therapeutic targets. FGFR2 serves as a model for studying receptor tyrosine kinases, a broader class of proteins frequently implicated in oncogenesis, thereby extending the relevance of FGFR2 research to other similar proteins [15].

Accurately linking sequence mutations in FGFR2 to functional outcomes is essential for advancing precision medicine, where treatments are tailored based on an individual's genetic makeup. This makes the study of FGFR2 both scientifically intriguing and clinically relevant. By developing methods that better capture the dynamic structural changes resulting from mutations in FGFR2, we can improve patient outcomes through more effective and personalized therapeutic strategies [6]. Ultimately, the insights gained from FGFR2 studies can have far-reaching implications, enhancing our understanding of protein function in health and disease and contributing to the development of innovative treatments.

In this paper, we develop a comprehensive workflow for mutant modeling by combining homology modeling methods such as SWISS-MODEL, AI-driven approaches like AlphaFold2 and AlphaMissense, and molecular modeling tools NAMD2 and SCWRL4. Rather than simply using these tools in isolation, we designed and tested multiple workflows by varying the number of rounds of NAMD2 and SCWRL4 refinements to identify the most reliable approach. Specifically, we evaluated different sequences and combinations of these tools to optimize both structural accuracy and computational efficiency. Our methodology aims to establish a robust framework for accurately predicting the effects of mutations in FGFR2, with rigorous evaluation metrics such as RMSD and TM score providing valuable information on the strengths and limitations of each protocol.

The methodology we present goes beyond the conventional use of individual modeling tools by systematically exploring various combinations and refinements through different workflows. By experimenting with multiple rounds of NAMD2 and SCWRL4, we have designed a flexible and dynamic approach that allows us to fine-tune structural predictions with unprecedented precision. This method is particularly compelling because it not only enhances the accuracy of FGFR2 mutant modeling but also provides a deeper understanding of the interplay between sequence variations and structural dynamics. The ability to test and optimize different protocols ensures that our workflow is adaptable and robust, making it a valuable tool for tackling the complex challenges of protein structure prediction. This approach offers a reliable pathway to link genetic mutations with functional outcomes, which is critical for the development of targeted therapies and precision medicine. The insights gained from this research have the potential to have an impact on a wide range of applications in structural bioinformatics and beyond.

In particular, our analysis reveals that the most effective approach for the modeling of mutants FGFR2 is achieved by combining AI-driven structure prediction methods with iterative rounds of molecular dynamics simulations and side-chain conformation optimization. Specifically, the workflow that integrates AlphaFold2 for initial structure prediction, followed by SCWRL4 refinement and short NAMD2 simulations, consistently produces the most accurate models, as evidenced by superior RMSD and TM-score values. This protocol, highlighted as Protocol C in Sect. 3, shows the highest precision in predicting the real structural status of the SNV FGFR2 models, significantly outperforming other workflows tested. The findings underscore the importance of extensive refinement to capture subtle yet critical structural variations that are essential for understanding the functional impact of mutations. Our results not only validate the proposed methodology, but also establish a clear protocol to achieve reliable and precise mutant modeling in FGFR2.

As we relate in Sect. 4, our findings reveal critical insights into the structural dynamics of FGFR2 and how single-nucleotide variants (SNVs) can lead to significant alterations in its function. We discover that certain mutations within the kinase domain of

FGFR2, when accurately modeled using our optimized workflow, result in subtle yet profound shifts in the protein's conformation that are not captured by less refined methods. These shifts are closely associated with changes in the ability of FGFR2 to interact with key signaling molecules, which may directly contribute to aberrant signaling pathways observed in various cancers and developmental disorders. By advancing our understanding of these mutation-induced structural changes, our study not only enhances the accuracy of FGFR2 mutant modeling but also provides valuable knowledge about the molecular mechanisms underlying diseases related to FGFR2. This contribution opens new avenues for targeted therapeutic interventions, reinforcing the importance of precise structural predictions in the context of personalized medicine.

2 Related Work

Recent advancements in protein structure prediction have significantly enhanced our understanding of how genetic mutations impact protein function, particularly in the context of disease [7, 16]. Among these advancements, AlphaFold2 has emerged as a groundbreaking tool, leveraging deep learning to predict protein structures with remarkable accuracy. AlphaFold2 uses multiple sequence alignments (MSA) to generate highly accurate 3D models, often rivaling experimental methods. Despite its success, AlphaFold2's limitations are notable when it comes to modeling the effects of single nucleotide variants (SNVs), particularly in specific proteins such as FGFR2. AlphaFold2 tends to maintain a bias toward an average, compact structure, which can obscure subtle but critical conformational changes induced by mutations.

AlphaMissense, a related AI-driven tool, addresses some of these limitations by focusing on the functional consequences of sequence variations. While it still relies on AlphaFold2 for its structural predictions, AlphaMissense integrates additional layers of analysis to predict how mutations may affect protein function. However, like AlphaFold2, it assumes that the structural predictions are sensitive to sequence mutations, which may not always be the case, especially for proteins like FGFR2 that exhibit complex dynamics and structural variability.

FGFR2 (Fibroblast Growth Factor Receptor 2) is a member of the receptor tyrosine kinase (RTK) family, playing a crucial role in cellular processes such as proliferation, differentiation, and survival. Mutations in FGFR2 are associated with various disorders, including craniosynostosis syndromes and several types of cancer. The FGFR2 tyrosine kinase domain (FGFR2-TKD) is particularly significant, as it mediates the intracellular signaling of the receptor and is frequently mutated in disease states.

Understanding the structural dynamics of FGFR2-TKD is essential to elucidate the mechanisms by which these mutations lead to disease. Traditional methodologies, such as homology modeling, have been widely used to predict the structure of FGFR2-TKD variants. Tools like SWISS-MODEL provide initial models based on known structures, offering a starting point for further refinement. However, these methods often fail to capture the dynamic nature of FGFR2, particularly the transitions between active and inactive states that are critical to its function.

Recent studies emphasize the importance of incorporating molecular dynamics (MD) simulations and side chain conformation optimization into the modeling workflow [17–20]. By using tools like NAMD2 and SCWRL4, researchers can refine the initial models generated by homology modeling or AI-driven predictions, allowing for a more accurate representation of the protein's dynamic behavior. For FGFR2-TKD, this is particularly important as mutations can induce significant conformational changes that affect the receptor's activity and its interaction with other molecules.

In this paper, we develop a comprehensive workflow that integrates these methodologies to address the limitations of existing work. By systematically varying the number of refinement rounds with NAMD2 and SCWRL4, we identify the most reliable approach for FGFR2 mutant modeling. Our results show that combining AlphaFold2 for initial structure prediction with extensive SCWRL4 and NAMD2 refinements produces the most accurate models, capturing the critical structural changes induced by mutations.

Moreover, our work advances knowledge about FGFR2 by providing a deeper understanding of its structural dynamics and how specific mutations impact its function. We have shown that certain mutations within the kinase domain of FGFR2, particularly those affecting the activation loop and kinase hinge regions, lead to conformational changes that are not adequately captured by traditional or AI-driven methods alone. These findings have significant implications for the development of targeted therapies, as they highlight the importance of considering both static and dynamic structural information when modeling protein function.

In summary, while tools such as AlphaFold2 and AlphaMissense represent significant advances in protein structure prediction, their limitations, particularly in the context of specific proteins like FGFR2, necessitate the development of more comprehensive modeling approaches. Our work contributes to this effort by combining multiple methodologies to create a robust workflow that not only improves the accuracy of FGFR2 mutant modeling but also enhances our understanding of the structural basis for FGFR2-related diseases.

3 Methods

3.1 Overview

This section presents the methodology for developing and evaluating our proposed workflow for FGFR2 mutant modeling. We integrate a range of computational approaches, including homology modeling (SWISS-MODEL), AI-driven structure prediction (AlphaFold2 and AlphaMissense), molecular dynamics (MD) simulations (NAMD2), and side-chain conformation optimization (SCWRL4) [21]. The objective is to establish a reliable and accurate approach to predict the structural impacts of single nucleotide variants (SNVs) in FGFR2, specifically within the tyrosine kinase domain (FGFR2-TKD). By systematically varying the number of refinement rounds, we identify the optimal sequence and combination of these tools to create a robust mutant modeling workflow.

3.2 Initial Structural Prediction

In addition to state-of-the-art methods, such as AlphaFold, we include in our evaluation classes, homology-based methods, such as SWISS-MODEL to obtain a comprehensive picture that is not skewed by the recent methods alone.

Homology Modeling with SWISS-MODEL. The initial structural models of FGFR2-TKD and its variants are generated using SWISS-MODEL, a homology modeling tool that constructs 3D structures based on sequence alignments with known templates from the Protein Data Bank (PDB). SWISS-MODEL is chosen for its ability to provide reliable structural templates that serve as the foundation for subsequent refinements. The sequences of the FGFR2 variants are aligned with the most homologous structures in the PDB, and the resulting models are used as starting points for further computational analyses.

AI-Driven Structure Prediction with AlphaFold2 and AlphaMissense. To supplement the homology-based models, we employ AlphaFold2 and AlphaMissense, which leverage deep learning techniques to predict protein structures from sequence data. AlphaFold2 uses multiple sequence alignments (MSA) to generate highly accurate structural models, while AlphaMissense focuses on predicting the effects of mutations on protein function. These AI-driven models provide an additional layer of structural prediction, enabling a comparison of traditional and cutting-edge approaches. The models generated by these tools are also subjected to further refinement to assess their utility in the modeling of FGFR2 mutants.

3.3 Local and Global Modeling of Dynamics: From Side-Chain Optimization to Whole Structure Dynamics

We consider both local and global structure dynamics, such as side-chain optimization through a representative, popular methods such as SCWRL4, and whole structure dynamics modeling through NAMD2.

Refinement with NAMD2. Molecular dynamics (MD) simulations are performed using NAMD2 to refine the initial structural models generated by SWISS-MODEL, AlphaFold2, and AlphaMissense. NAMD2 is employed to explore the conformational space of FGFR2-TKD and its variants, allowing for the relaxation and adjustment of the initial models to more accurately reflect their dynamic behavior. The CHARMM force field is used to simulate the physical interactions within the protein, and the simulations are run for varying lengths of time, depending on the complexity of the structure being modeled. The refinement process aims to reduce the Root Mean Square Deviation (RMSD) between the predicted models and the experimentally resolved structures, thus improving the accuracy of the structural predictions.

Side-Chain Conformation Optimization with SCWRL4. Following MD simulations, SCWRL4 is used to optimize the side-chain conformations of the refined FGFR2 models. SCWRL4 uses a backbone-dependent rotamer library to predict the most likely positions of amino acid side chains, which is crucial to accurately representing the properties and interactions of the protein surface. The side-chain optimization process is iterated multiple times to ensure convergence to the most stable conformation. This step is particularly important for understanding how mutations in FGFR2 impact its structural integrity and function, as side-chain interactions often play a critical role in protein stability and activity.

3.4 Workflow Design and Evaluation

To identify the most effective workflow for the modeling of FGFR2 mutants, we design and test several workflows by varying the number of refinement rounds for NAMD2 and SCWRL4. Three main workflows are established and compared: Workflow A involves a single round of NAMD2 followed by SCWRL4; Workflow B involves 50 rounds of NAMD2 and 50 rounds of SCWRL4; and Workflow C reverses the order of Workflow B, applying 50 rounds of SCWRL4 before NAMD2.

The selection of 50 refinement rounds was guided by empirical analysis, which indicated diminishing returns beyond this threshold. Preliminary tests showed that RMSD and TM-score improvements became marginal after 50 iterations, while computational costs increased significantly. Additionally, prior studies suggest that similar iterative refinements stabilize after a finite number of cycles, supporting our decision to limit the number of rounds to 50 as a practical trade-off between accuracy and efficiency.

Each protocol is applied to model 17 SNV FGFR2 structures, selected based on their known pathogenic relevance and diversity across key functional regions, such as the activation loop, kinase hinge, and catalytic pocket. It is important to clarify that earlier workflow optimizations were performed on a subset of SNVs (e.g., 13 or 15), while the final structural evaluations were conducted on 17 SNV models. This explains the variation in SNV counts referenced across different sections of this paper.

The effectiveness of each workflow is evaluated using two key metrics: Root Mean Square Deviation (RMSD) and TM-score. RMSD measures the average distance between atoms of the superimposed structures, providing a quantitative assessment of the accuracy of the predicted models. A lower RMSD indicates a closer match to the reference structure. The TM-score, on the other hand, evaluates the global similarity of the protein folds, offering a complementary perspective on model accuracy. Scores closer to 1 indicate a higher structural similarity, with TM-scores above 0.5 generally considered to represent accurate structural predictions. Both metrics are calculated for each FGFR2 variant under the different workflow protocols, allowing us to identify the combination of steps that produces the most accurate and reliable structural models.

As related in Sect. 4, once a workflow is determined as more precise for modeling structure, its findings are then utilized to link sequence to function. Specifically, we employ the following approach to classify variants based on the functional regions they affect and the potential disease classification. To categorize the predicted structures of FGFR2 SNVs, we use Principal Component Analysis (PCA), a dimensionality reduction technique that transforms complex structural data into principal components, capturing the most significant variance. By projecting the tertiary structures of wild-type and variant FGFR2 models onto the top two principal components, we are able to visually and quantitatively assess the structural differences and similarities among the variants.

We are able to distinguish between key functional regions, such as the activation loop, kinase hinge, and nucleotide-binding loop, which are crucial for FGFR2's activity [12, 13, 22]. Variants with mutations in these regions showed distinct clustering patterns, indicating how these structural changes may impact the protein's function. The PCA analysis provided a clear visualization of how specific mutations alter the structural landscape of FGFR2, contributing to our understanding of their functional implications.

In addition to the classification of functional regions, we can also reveal the relationship between structural variations and disease phenotypes. By overlaying disease classifications onto the PCA projections, we identify distinct clusters corresponding to different diseases, such as craniosynostosis syndromes and cancers. This classification highlights the structural basis for the disease-specific effects of FGFR2 mutations, offering valuable information on how these mutations lead to specific pathological outcomes.

4 Results

Summary of Key Findings. The comparative analysis related in greater detail below reveals that Protocol C, which involved 50 rounds of SCWRL4 followed by 50 rounds of NAMD2, consistently produced the most accurate models, as indicated by the lowest RMSD and highest TM scores. This protocol effectively captures the critical structural changes induced by FGFR2 mutations, particularly in regions such as the activation loop and the kinase hinge, which are crucial for receptor function. Protocol B, while also effective, is more computationally intensive, and Protocol A, with only a single round of refinement, is insufficient for complex structures like FGFR2-TKD. The results of this analysis underscore the importance of extensive refinement in mutant modeling workflows, particularly for proteins like FGFR2 that exhibit complex structural dynamics. By combining AI-driven predictions with thorough molecular dynamics simulations and side-chain optimization, we are able to achieve a level of accuracy that surpasses traditional methods. These findings suggest that the proposed workflow, particularly Protocol C, can serve as a reliable framework for modeling the effects of SNVs in FGFR2 and potentially other proteins with similar structural characteristics.

Data. In our study, we utilize a dataset comprising 7 wild type, 17 single-nucleotide variants (SNVs) of FGFR2, 5 multiple-nucleotide variants (see Table 1), all of which are obtained from the Protein Data Bank (PDB). These variants are mapped to specific functional regions within the FGFR2 tyrosine kinase domain (FGFR2-TKD), each playing a crucial role in the protein's structural and functional dynamics. The residues 525–539, the gatekeeper residue (position 564), the kinase hinge region (residues 566–571, 549, 565), the kinase insert (residues 579–599), the catalytic pocket (residues 620–630), the Alpha-C tether (residue 650) and the activation loop (residues 643–649, 651–664 [12, 13, 22].

Each mutation is analyzed to determine its impact on these functional regions, and the associated disease phenotypes are identified. Table 1 categorizes these mutations by their positions, indicating the functional regions affected and the corresponding diseases linked to each mutation.

In addition to these resolved structures, our dataset includes 6 unresolved FGFR2 protein sequences. These sequences are critical for our subsequent modeling efforts. Among these, some are associated with benign conditions, while others are linked to pathogenic mutations. We will use these sequences to further investigate and model their structural impacts, particularly in relation to the functional dynamics of FGFR2.

4.1 Evaluation of Initial Structural Models for SNV FGFR2 Structures

The initial structural models for 15 single-nucleotide variant (SNV) FGFR2 structures are generated using SWISS-MODEL, AlphaFold2, and AlphaMissense, and evaluated for structural accuracy using RMSD and TM-score metrics (see Table 2). The models generated by SWISS-MODEL show an average RMSD of 2.1 Å and a TM-score of 0.85, indicating moderate accuracy. AlphaFold2 demonstrates superior accuracy with an average RMSD of 1.3 Å and a TM-score of 0.92, indicating a high level of structural similarity to the reference structures. Similarly, AlphaMissense produces initial models with an average RMSD of 1.4 Å and a TM-score of 0.90, reflecting high structural accuracy. These results highlight that AI-driven methods, particularly AlphaFold2, yield more accurate initial structural models compared to the homology modeling approach of SWISS-MODEL, underscoring the importance of choosing the appropriate modeling tool based on the desired level of accuracy for initial structural predictions (see Figs. 1, 2).

Table 1. This table summarizes the mutations in various FGFR2 structures, classified by their PDB ID or Unresolved Sequence, the specific mutation, the region name where the mutation occurs, the functional status of the protein (active, inactive, or unclear), and the associated disease.

PDB ID	Mutation Position	Region Name	Status	Disease
1GJO	WT	-	Inactive	-
1OEC	WT	-	Inactive	-
2PSQ	WT	-	Inactive	-
2PVF	WT	-	Active	-
3RI1	WT	-	Inactive	-
6LVK	WT	-	Unclear	-
6LVL	WT	-	Unclear	-
2PVY	K659N	Activation-Loop	Active	UCS
2PWL	N549H	Kinase Insert	Active	CS
2PY3	E565G	Kinase Insert	Active	PS
2PZ5	N549T	Kinase Insert	Active	PS
2PZP	K526E	αC helix at N	Active	CS
2PZR	K641R	Kinase Insert	Active	PS
2Q0B	E565A	Kinase Insert	Active	PS
3B2T	A628T	Catalytic Pocket	Active	LADD1
3CLY	C491A	-	Unclear	-
4J95	K659N	A-Loop	Active	UCS

(*continued*)

Table 1. (*continued*)

PDB ID	Mutation Position	Region Name	Status	Disease
4J96	K659M	A-Loop	Active	CC
4J97	K659E	A-Loop	Active	EC
4J98	K659Q	A-Loop	Active	-
4J99	K659T	A-Loop	Active	-
5UGL	D650V	αC tether	Active	-
7KIA	V564F	Gate Keeper	Unclear	-
7KIE	V564F	Gate Keeper	Unclear	-
5EG3	Multiple	A-Loop	Active	-
5UGX	E565A/D650V	KH/αC tether	Active	PS
5UHN	E565A/N549H	KH	Active	PS/CS
5UI0	E565A/K659M	KH/A-Loop	Active	PS/CC
6V6Q	Multiple	-	Unclear	-
Unresolved Sequence	R612T	-	Unclear	LC
Unresolved Sequence	A648T	A-Loop	Unclear	LADD1
Unresolved Sequence	G663E	A-Loop	Unclear	PS
Unresolved Sequence	R678G	-	Unclear	CS
Unresolved Sequence	Q597K	Kinase Insert	Unclear	-
Unresolved Sequence	I654L	A-loop	Unclear	-

Abbreviations: KH = Kinase Hinge; UCS = Unicoronal Craniosynostosis; CS = Crouzon Syndrome; PS = Pfeiffer Syndrome; LADD1 = Lacrimo-auriculo-dento-digital syndrome; LC = Lung Cancer.

Table 2. This table presents the structural accuracy metrics for initial models of 15 single-nucleotide variant (SNV) FGFR2 structures generated using SWISS-MODEL, AlphaFold2, and AlphaMissense. The metrics used for evaluation are Root Mean Square Deviation (RMSD) and TM-score, averaged across the 15 structure for each method.

	RMSD (Å)	TM-score
SWISS-MODEL	2.1	0.85
AlphaFold2	1.3	0.92
AlphaMissense	1.4	0.90

4.2 Refinement and Optimization of FGFR2 Models

The application of NAMD2 for molecular dynamics (MD) simulations significantly improved the structural accuracy of the initial models (see Table 3). For SWISS-MODEL generated structures, the average RMSD decreased to 1.8 Å, and the TM-score improved to 0.88. Refinement of AlphaFold2 models with NAMD2 resulted in an average RMSD of 1.1 Å and a TM-score of 0.93, while AlphaMissense models showed an RMSD of 1.2 Å and a TM-score of 0.91 after refinement.

Further optimization with SCWRL4, which adjusts side-chain conformations, provided additional enhancements. For SWISS-MODEL, the final models achieved an RMSD of 1.7 Å and a TM-score of 0.89. AlphaFold2 models refined with SCWRL4 reached an RMSD of 1.0 Å and a TM-score of 0.94, and AlphaMissense models achieved an RMSD of 1.1 Å and a TM-score of 0.92.

Table 3. This table shows the structural accuracy metrics after refinement of the initial models for 15 SNV FGFR2 structures generated by SWISS-MODEL, AlphaFold2, and AlphaMissense. The refinement was done using NAMD2 for molecular dynamics simulations and SCWRL4 for side-chain conformation optimization. The metrics used for evaluation are Root Mean Square Deviation (RMSD) and TM-score, averaged across the 15 structures for each method.

Method	Refinement Step	Average RMSD (Å)/TM-score
SWISS-MODEL	NAMD2	1.8/0.88
SWISS-MODEL	NAMD2 + SCWRL4	1.7/0.89
AlphaFold2	NAMD2	1.1/0.93
AlphaFold2	NAMD2 + SCWRL4	1.0/0.94
AlphaMissense	NAMD2	1.2/0.91
AlphaMissense	NAMD2 + SCWRL4	1.1/0.92

4.3 Comparative Analysis of Workflow Effectiveness

To determine the most reliable approach for structural modeling of FGFR2, we design and test three different modeling workflows using a blind test strategy. The first protocol involved initial structural modeling followed by a single round of refinement with NAMD2 and SCWRL4. The second protocol consists of the structural modeling step followed by 50 rounds of NAMD2 refinement and 50 rounds of SCWRL4 refinement. The third protocol applies 50 rounds of SCWRL4 refinement first, followed by 50 rounds of NAMD2 refinement.

These sequences are applied to the 15 SNV FGFR2 models using both inactive and active reference structures as templates. The resulting models undergo further refinement with NAMD2 and SCWRL4, and their energy minimization is assessed. Models with minimized energy levels closer to the inactive reference are classified as inactive, whereas those closer to the active reference are classified as active.

The structural states predicted by each modeling sequence are then compared to the known real structural statuses of the SNVs, allowing us to evaluate the reliability of each sequence in accurately predicting the functional state (active or inactive) of the modeled structures. The first protocol shows moderate accuracy in predicting the actual structural

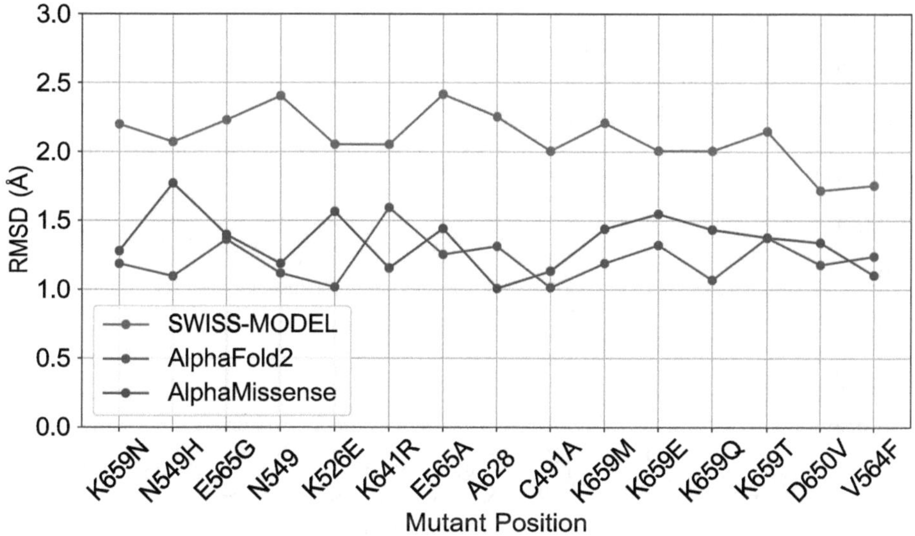

Fig. 1. Fifteen mutant positions of FGFR2-TKD are evaluated using three structural prediction methods: SWISS-MODEL, AlphaFold2, and AlphaMissense. The RMSD values, which measure the average deviation between the predicted and reference structures, reveal significant differences in accuracy. SWISS-MODEL shows the highest RMSD values, averaging around 2.1 Å, indicating a larger deviation from the reference structure. In contrast, AlphaFold2 achieves the lowest RMSD values, with an average of approximately 1.3 Å, suggesting superior accuracy in predicting the protein structure. AlphaMissense also performs well, with an average RMSD of about 1.4 Å, closely following AlphaFold2 in terms of accuracy.

status, but the single round of refinement is often insufficient for more complex structures. The second protocol improved accuracy with extensive refinement cycles, leading to better energy minimization and more reliable state predictions, though it is computationally intensive. The third sequence provides the best balance between computational efficiency and accuracy. By applying extensive SCWRL4 refinement before NAMD2, this approach effectively captures side-chain conformations and preserved overall structural integrity, resulting in the highest agreement with the real structural statuses of the SNVs.

In general, the third modeling workflow (see Table 4) proved to be the most reliable for predicting the actual structural status of the SNV FGFR2 models, providing valuable insights for future mutant modeling studies specific for FGFR2.

4.4 PCA Classification of FGFR2 Variants Using the Optimal Workflow

To further assess the effectiveness of the optimal workflow (SCWRL4 + NAMD2) combined with AlphaFold2, we performed a Principal Component Analysis (PCA) on the predicted structures of FGFR2 SNVs. This analysis was designed to classify the variants based on their structural characteristics and to explore how these features correlate with functional regions and disease states. Specifically, we selected 6 SNVs, including 2

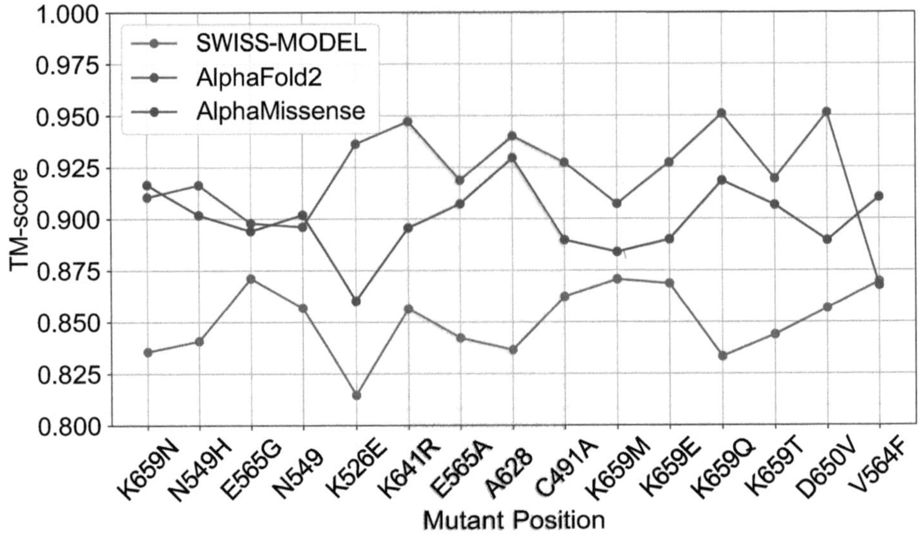

Fig. 2. The same fifteen mutant positions of FGFR2-TKD are further assessed through TM-score comparisons across the three prediction methods. TM-score evaluates the structural similarity between the predicted and reference structures, with higher scores indicating closer resemblance. AlphaFold2 achieves the highest TM-score, averaging around 0.92, demonstrating the most accurate structural alignment with the reference. AlphaMissense follows closely with an average TM-score of approximately 0.90, also reflecting strong structural similarity. SWISS-MODEL, while still producing reasonably high TM-scores with an average of 0.85, shows slightly less consistency in matching the reference structure compared to the other two methods.

benign and 4 pathogenic mutations (see Table 1), and subjected them to mutant modeling using our optimized workflow. These newly modeled structures were then projected onto the principal components derived from the PCA of 53 known FGFR2 structures sourced from the PDB.

Functional Region Classification. The PCA embedding allowed us to project the tertiary structures of the wild-type and variant FGFR2 models onto the top two principal components, effectively capturing the main sources of structural variation. Key functional regions, such as the activation loop, kinase hinge, and nucleotide-binding loop, were prominently represented in the PCA results, reflecting their critical roles in FGFR2 activity. The SNVs with mutations in these regions displayed distinct clustering patterns, underscoring the structural impact of specific mutations on the protein's function. The projection of the 6 modeled SNVs onto this PCA space provided additional insights, revealing how these newly studied variants align with established structural variations, further validating the robustness of our workflow.

The figure (see Fig. 3) presents a detailed analysis of the structural dynamics of the FGFR2 tyrosine kinase domain (FGFR2-TKD), focusing on how mutations impact the activation states of the protein. The figure compares the positions of different mutant and wildtype structures in a reduced 2D space derived from principal component analysis

Table 4. This table summarizes the accuracy of various structural modeling methods when combined with three different modeling protocols in predicting the actual structural status (active or inactive) of 15 SNV FGFR2 structures. Accuracy is measured as the percentage of correct predictions relative to the known real statuses of the structures.

Structural Modeling Method + Protocol	Accuracy (%)
SWISS-MODEL + Protocol A	74.32
SWISS-MODEL + Protocol B	82.35
SWISS-MODEL + Protocol C	82.35
AlphaFold2 + Protocol A	75.79
AlphaFold2 + Protocol B	88.24
AlphaFold2 + Protocol C	88.24
AlphaMissense + Protocol A	72.31
AlphaMissense + Protocol B	82.35
AlphaMissense + Protocols C	88.12

(PCA), highlighting the structural differences among various activation states. The figure shows that mutations, particularly those in the activation loop (such as K659N, K659M, and K659E), result in a clustering of these variants in a similar region of the 2D space, indicating similar structural dynamics. These mutations are associated with increased kinase activity, a common feature in oncogenic variants. Other mutations, like those in the kinase hinge region, are also plotted to demonstrate their distinct positioning, suggesting different functional impacts.

Disease Classification. The figure (see Fig. 4) presents a refined principal component analysis (PCA) embedding of FGFR2 tyrosine kinase domain (FGFR2-TKD) structures, integrating both known and computationally predicted mutant models. The figure explores the relationship between structural dynamics and disease classification by projecting these structures into a 2D space and color-coding them according to their associated disorders. Mutations linked to genetic and developmental disorders, such as Crouzon syndrome (R678G) and Pfeiffer syndrome (G663E), show a clear tendency to cluster together, reflecting similar structural impacts. In contrast, mutations associated with cancers, such as cervical cancer (K659M) and lung cancer (R612T), are positioned distinctly, suggesting divergent structural alterations that correlate with oncogenic activity. The inclusion of computationally modeled structures, such as A648T and R612T, further enriches the analysis by extending the observable patterns to mutations with no previously resolved structures. These models align closely with their respective disease clusters, confirming their predicted pathogenic roles. The separation between developmental and oncogenic mutations in the PCA space highlights the structural basis for the functional and clinical differences observed in FGFR2-related disorders.

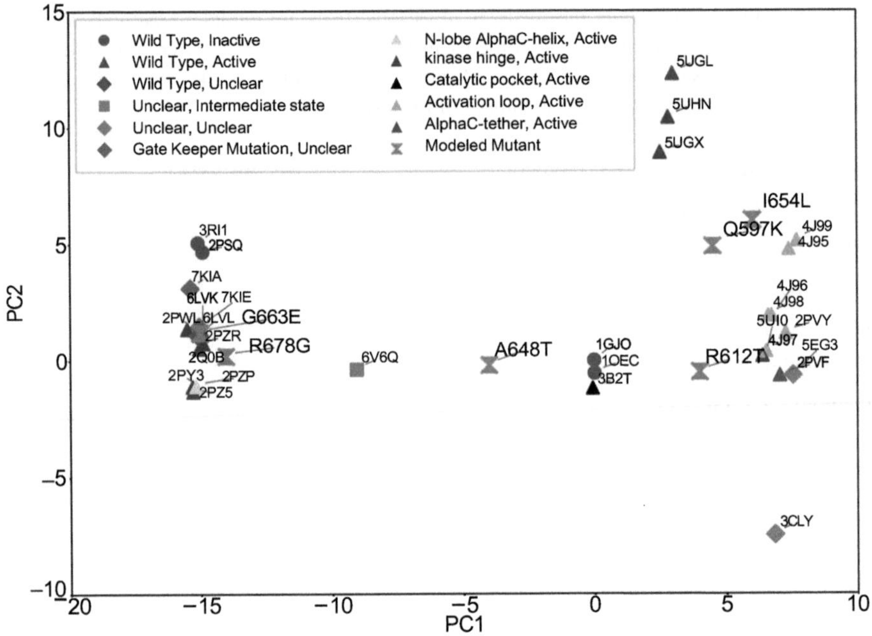

Fig. 3. PC1-PC2 embedding of known and computed structures. Colors indicate functional regions, and shapes indicate state. 'X' is reserved to indicate the computed structures.

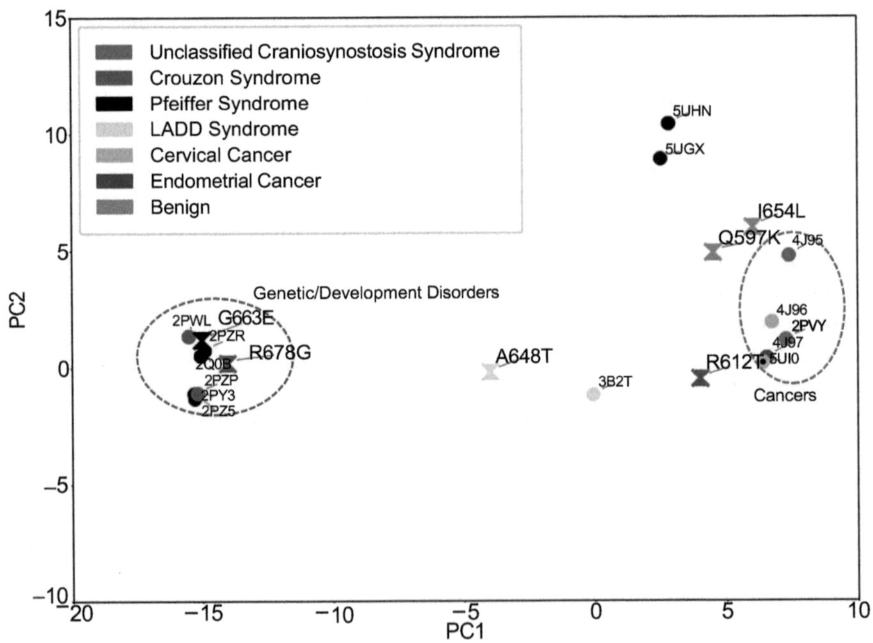

Fig. 4. PC1-PC2 embedding of known and computed structures. Color-coding indicates disease classifications.

5 Conclusion

This study presents a novel and comprehensive approach to modeling FGFR2 mutations, leveraging state-of-the-art computational tools to achieve unprecedented accuracy in structural predictions. By integrating AI-driven methods like AlphaFold2 and AlphaMissense with molecular dynamics simulations (NAMD2) and side-chain conformation optimization (SCWRL4), we developed a workflow that effectively captures the intricate structural dynamics induced by single-nucleotide variants (SNVs) within the FGFR2 tyrosine kinase domain. Our findings show that the optimized protocol, particularly the sequence involving 50 iterative rounds of SCWRL4 refinement followed by NAMD2 simulations, consistently outperforms traditional approaches in predicting both the structural configurations and functional consequences of FGFR2 mutations.

The significance of this work lies in its ability to incorporate dynamic information into the modeling process, which is crucial for understanding how specific mutations impact FGFR2's function. Our analysis revealed that mutations in key regions, such as the activation loop and kinase hinge, result in conformational changes that are strongly correlated with altered kinase activity and the onset of diseases, including various cancers and developmental disorders. These insights are critical for elucidating the molecular mechanisms underlying FGFR2-related diseases and have the potential to inform the development of more effective, targeted therapeutic interventions.

Beyond FGFR2, the methodology developed in this study has broad implications for the field of structural bioinformatics. It provides a reliable and adaptable framework for studying the effects of genetic mutations on a wide range of proteins, particularly those with complex structural and functional characteristics. As such, this approach could play a pivotal role in advancing precision medicine by enabling more accurate predictions of mutation-driven disease mechanisms and supporting the design of personalized treatment strategies. Future research could further refine this workflow, explore its applicability to other challenging protein targets, and expand its use in drug discovery and development.

References

1. Saraon, P., Pathmanathan, S., Snider, J., Lyakisheva, A., Wong, V., Stagljar, I.: Receptor tyrosine kinases and cancer: oncogenic mechanisms and therapeutic approaches. Oncogene **40**(24), 4079–4093 (2021)
2. Balogun, T.A., et al.: Receptor tyrosine kinases as a therapeutic target by natural compounds in cancer treatment. Future J. Pharm. Sci. **7**(1), 1–15 (2021)
3. Glaser, R.L., et al.: Paternal origin of FGFR2 mutations in sporadic cases of Crouzon syndrome and Pfeiffer syndrome. Am. J. Hum. Genet. **66**(3), 768–777 (2000)
4. Wenger, T., Miller, D., Evans, K.: FGFR craniosynostosis syndromes overview. GeneReviews®[Internet] (2020)
5. Nakamura, I.T., et al.: Comprehensive functional evaluation of variants of fibroblast growth factor receptor genes in cancer. NPJ Precis. Oncol. **5**(1), 66 (2021)
6. Du, Z., Lovly, C.M.: Mechanisms of receptor tyrosine kinase activation in cancer. Mol. Cancer **17**, 1–13 (2018)
7. Casaletto, J.B., McClatchey, A.I.: Spatial regulation of receptor tyrosine kinases in development and cancer. Nat. Rev. Cancer **12**(6), 387–400 (2012)

8. Hevner, R.F.: Brain overgrowth in disorders of RTK–PI3K–AKT signaling: a mosaic of malformations. Semin. Perinatol. **39**, 36–43 (2015)
9. Waterhouse, A., Bertoni, M., Bienert, S., Studer, G., Tauriello, G., Gumienny, R., Heer, F.T., de Beer, T.A.P., Rempfer, C., Bordoli, L., et al.: Swiss-model: homology modelling of protein structures and complexes. Nucleic Acids Res. **46**(W1), W296–W303 (2018)
10. Cheng, J., et al.: Accurate proteome-wide missense variant effect prediction with alphamissense. Science **381**(6664), eadg7492 (2023)
11. Jumper, J., et al.: Highly accurate protein structure prediction with alphafold. Nature **596**(7873), 583–589 (2021)
12. Chen, H., et al.: Cracking the molecular origin of intrinsic tyrosine kinase activity through analysis of pathogenic gain-of-function mutations. Cell Rep. **4**(2), 376–384 (2013)
13. Chen, H., et al.: Elucidation of a four-site allosteric network in fibroblast growth factor receptor tyrosine kinases. eLife **6**, e21137 (2017)
14. Kalé,L., et al.: NAMD2: greater scalability for parallel molecular dynamics. J. Comput. Phys. **151**(1), 283–312 (1999)
15. Amadei, A., Linssen, A.B., Berendsen, H.J.: Essential dynamics of proteins. Proteins: Struct. Funct. Bioinform. **17**(4), 412–425 (1993)
16. Regad, T.: Targeting RTK signaling pathways in cancer. Cancers **7**(3), 1758–1784 (2015)
17. Lian, Y., Bodian, D., Shehu, A.: Elucidating the role of wildtype and variant FGFR2 structural dynamics in (Dys) function and disorder. Int. J. Mol. Sci. **25**(8), 4523 (2024)
18. Brown, L.M., Ekert, P.G., Fleuren, E.D.G.: Biological and clinical implications of FGFR aberrations in paediatric and young adult cancers. Oncogene **42**, 1875–1888 (2023)
19. Karasev, D., Sobolev, B., Lagunin, A., Filimonov, D., Poroikov, V.: Predicting the protein-ligand interactions based on ligands' structures and proteins' sequences. In: Bioinformatics of Genome Regulation and Structure/Systems Biology (BGRS/SB-2022), pp. 345–345 (2022)
20. Jing, X., Wu, F., Luo, X., Xu, J.: Raptorx-single: single-sequence protein structure prediction by integrating protein language models. bioRxiv, pp. 2023-04 (2023)
21. Krivov, G.G., Shapovalov, M.V., Dunbrack Jr, R.L.: Improved prediction of protein side-chain conformations with scwrl4. Proteins: Struct. Funct. Bioinform. **77**(4), 778–795 (2009)
22. Chen, H., et al.: A molecular brake in the kinase hinge region regulates the activity of receptor tyrosine kinases. Mol. Cell **27**(5), 717–730 (2007)

A Novel Dataset and Comprehensive AI Model Evaluation for Masseter Muscle Segmentation

Yunfeng Liang[1(✉)], Lin Zou[1], Millie Ming Rong Goh[1], Si Xian Ho[1], Zhen Jie Zhang[1], Barbara Helen Rosario[2,3], Bharathi Balasundaram[3,4,5], Sindhu John[3,6], Saraswathy Suresh Babu[3,6], Charlene Jin Yee Liew[3,6], Aileen Lim[7], Andy Wee An Ta[1], and Han Leong Goh[1]

[1] Data aNalytics and AI, Synapxe Pte Ltd, Singapore, Singapore
{liang.yunfeng,jenny.zou,millie.goh,ho.si.xian1,zhang.zhenjie, andy.ta,hanleong.goh}@synapxe.sg
[2] Department of Geriatric Medicine, Changi General Hospital, Singapore, Singapore
rosario.barbara.helen@singhealth.com.sg
[3] Duke-NUS Graduate Medical School, Singapore, Singapore
{balasundaram.bharathi,sindhu.john,saraswathy.suresh.babu, charlene.liew.j.y}@singhealth.com.sg
[4] Department of Psychological Medicine, Changi General Hospital, Singapore, Singapore
[5] NUS Yong Loo Lin School of Medicine, Singapore, Singapore
[6] Department of Diagnostic Radiology, Changi General Hospital, Singapore, Singapore
[7] Department of Health Services Intelligence, Changi General Hospital, Singapore, Singapore
aileen_lim@cgh.com.sg

Abstract. Accurate segmentation of the masseter muscles (MM) is critical for the assessment of various clinical conditions. While artificial intelligence (AI)-based methods have demonstrated promising results, existing studies suffer from limitations such as small datasets and constrained model deployment bandwidth, hindering cross-study comparisons. To address these challenges, we curated a comprehensive dataset comprising over 1,000 annotated images from four distinct sources. We trained and evaluated a diverse set of architectures including the well-applied convolutional neural network (CNN)-based models as well as the recently proposed Vision Transformer and Vision Mamba based models. A standardized evaluation pipeline was developed to assess various performance metrics across models. Our results demonstrate that the YOLOv8 + SAM model exhibits superior generalization capabilities compared to others.

Keywords: masseter muscle segmentation · deep learning · dataset curation

1 Introduction

The masseter muscles (MM) play a crucial role in the function and structure of the jaw [1]. The size and shape of the MM have a great impact on facial aesthetics, jaw health, and can be seen as a key indicator of physical frailty, cognitive decline and aging [3, 4]. Accurate segmentation of the MM is an important step for assessing the conditions mentioned above.

T. Aldwairi et al. (Eds.): BICOB 2025, CCIS 2535, pp. 85–95, 2025.
https://doi.org/10.1007/978-3-031-94039-2_7

Efforts have been made on the automation of the MM segmentation process by using computer technologies to relieve the enormous clinical effort of manual data labeling. Despite the early works of using the traditional image processing techniques [6–8], artificial intelligence (AI) technologies have drawn attention for their superior performance in recent years [9, 10] and several deep learning based methods have been developed for automated MM segmentation across various imaging modalities.

One of the most extensively studied models for medical image segmentation is U-Net [12] and its numerous variants [13, 14]. For instance, Orhan et al. [11] compared U-Net with the Pyramid Scene Parsing Network (PSPNet) and Fuzzy Petri Net (FPN) on 195 ultrasonography (USG) images from 24 patients, demonstrating promising results in MM segmentation and thickness measurement. Similarly, Keser et al. [17] reported effective performance using U-Net on 388 USG images. Jiang et al. [16] proposed a U^2-Net-based model for conebeam computed tomography (CBCT) segmentation, leveraging on high-quality CT data for refinement. Besides U-Net, other models such as residual networks [5], DeepLabv3 + [20], and graph reasoning techniques [21], have also been explored with encouraging results.

One common problem with the existing works is the lack of data. To the best knowledge of the authors, there is no publicly available dataset specifically designed for MM segmentation. Some existing datasets [22, 23] contain the head and neck region data needed for MM segmentation. However, they are designed for other tasks with a large portion of the datasets not relevant and all without the MM-specific annotations. Consequently, many studies rely on small selfcollected datasets, which constrains model performance and impedes cross-study comparisons.

Another notable issue is the predominant reliance on CNN-based architectures. While CNNs have been the backbone of image processing tasks over the past decade, emerging models such as Vision Transformers and Vision Mamba have demonstrated superior capabilities in capturing long-range dependencies, making them particularly well-suited for complex segmentation tasks.

Therefore, to address the above-mentioned challenges, we carried out this work with the following contributions. First, we created an assembled dataset specially designed for the MM segmentation task by collecting and labeling the data from variant sources. Different data modalities have been used for MM segmentation tasks and their differences have been discussed [16, 19]. We chose to use the CT image modality because its high quality is very suitable for this accuracy-oriented task and there are enough public data sources available to construct the new set as well. Second, we designed the pipeline of data analysis for model performance comparison. Third, we carried out experiments based on the designed pipeline using variant AI models, including the previously adopted CNN networks, as well as the newly developed transformer and mamba architectures. Variant evaluation metrics are inspected to have a detailed examination of different perspectives of the models.

The rest of the paper is organized as follows. Sect. 2 introduces the proposed data collection and model evaluation pipeline. The testing results on different datasets are presented in Sect. 3. Finally, the paper is concluded in Sect. 4 after the discussion on the results.

2 Methodology

2.1 Data Preparation

To collect a dataset that is large and diverse for proper training and comprehensive evaluation of the models, we have looked into different publicly available CT databases including CQ500 [22], SinoCT [24] and PDDCA [23]. The CQ500 dataset is collected from multiple radiology centers in the New Delhi region which contains anonymized CT scans of 491 patients. The data is collected using multiple scanners from GE or Philips with different settings. The pixel spacing in the axial plane ranges from 0.38 mm to 0.79 mm. The SinoCT is collected by the Center for Artificial Intelligence in Medicine & Imaging of Stanford University. It contains more than 9,000 head CT scans. The data is collected by the same GE scanner model with a pixel spacing of 0.49 or 0.60 mm in the axial plane. The PDDCA dataset is introduced in the MICCAI 2015 Head and Neck AutoSegmentation Challenge, which consists of 48 CT scans with an axial plane pixel spacing between 0.76 mm and 1.25 mm. In addition to the public available data, we also collected our private CT data from Changi General Hospital (CGH) in Singapore, which consists of CT scans of 14 patients with a pixel spacing of 0.51 mm in the axial plane. Due to the data privacy reason, the CGH data will not be shown in the paper and only the result statistics is discussed.

The first step of the data preparation is selecting suitable scan data for our specific topic. Because the public datasets mentioned above are not designed for the MM segmentation problem, a large portion of the collections do not have the corresponding region in the scan. The whole process of the data collection is shown in Fig. 1. First, we manually identified the CT scans that contain the facial area of the patients. After that, a further selection of the slices in the axial plane of each scan is carried out. All the slices that contain the whole MM regions are extracted and saved. The reconstructed images are 512 × 512 pixels for all the datasets. Data annotation is done by using the segment anything annotator tool. It proposes an initial segmentation mask in the region selected and allows further manual refinement.

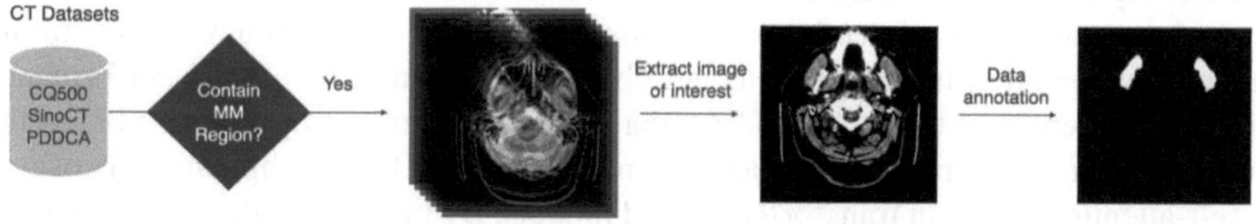

Fig. 1. Flow chart of the data collection process.

The curated dataset details and the corresponding extraction summaries are presented in Table 1. Selected CT slices and annotation pairs are later used in model training. Representative image examples and the corresponding ground truth labelings are shown in Fig. 2. Notably, the PDDCA dataset exhibits distinct characteristics due to the larger pixel spacing values and smaller region of interest.

The collected dataset is split into two groups, the images from CQ500 and SinoCT are combined and used for both model training and testing while the ones from PDDCA

Table 1. Summary of the collected dataset.

Source	Format	Patient #	Image #	Pixel Spacing	Usage
CQ500 [22]	DICOM	162	413	0.38–0.79 mm	Train & Test
SinoCT [24]	DICOM	499	825	0.49 mm	Train & Test
CGH	DICOM	14	14	0.51 mm	Test
PDDCA [23]	NRRD	30	30	0.76–1.25 mm	Test

and CGH are only used for testing. This aims to test the generalization capability of the models by using the trained models on data that collected from different environment and equipment, which is very critical in putting the AI models into real usage. For the model training, 80–20 train-test split of the CQ500 + SinoCT data is adopted and 20% of the training is held out as the validation set. Additionally, all the data splitting is done at the patient level to avoid data leakage.

2.2 Evaluation Pipeline

Based on the image dataset prepared, an end-to-end pipeline is designed to train and evaluate different AI models altogether.

Models. The models being tested includes DeepLabv3+ [25], U^2-Net [13], nnUNet [14], YOLOv8 + SAM [27, 28], and U-Mamba [26]. The DeepLabv3 + model is designed to extract multi-scale context information by applying dilated convolutions with different rates. U^2-Net is a lightweight yet highly effective model using the nested U-Net structure for salient object detection. Both models belong to CNN and have already been applied in the MM segmentation problem as reviewed in the introduction. The nnU-Net is another variant U-Net that has achieved good performance in many medical image segmentation tasks [29]. The YOLOv8 + SAM method is a two-stage approach that first detects bounding boxes for the areas of interest and then feeds them as prompts together with the original image to SAM for mask generation. The integration combines YOLO's fast and accurate object detection capabilities with SAM's proficiency in generating detailed segmentation masks for detected objects, delivering exceptional performance in image segmentation tasks. The U-Mamba model is recently proposed to address the issue of handling long-range dependencies in CNN and the high computational costs of the self-attention layers in transformer-based models. By integrating the Mamba blocks into the U-Net architecture, the U-Mamba is shown to have better performance than other models across various medical image segmentation tasks.

The detailed settings of model training are listed in Table 2. All models except U^2-Net and SAM can take the original images as input without any scaling. The input for U^2-Net is smaller than the image collected and linear interpolation is adopted in training and output evaluation to resize the images to 320 x 320. On the other hand, the input for SAM is larger than the image collected and the cubic interpolation is used to upscale the image to 1024 x 1024. The pixel spacing values in both models are also scaled accordingly for the computation of distance-based metrics. The optimization and data augmentation are performed based on the original paper. By default, nnU-Net and U-Mamba do not have

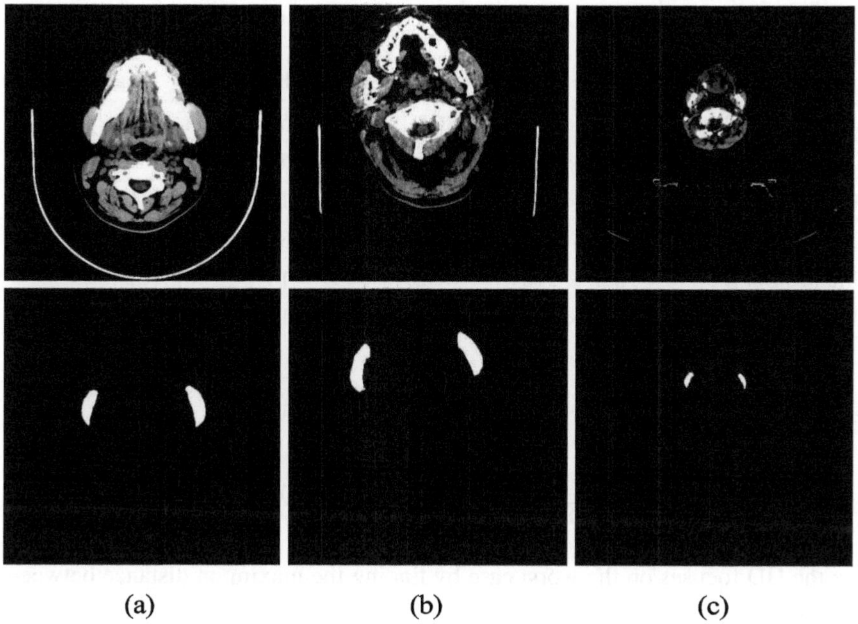

Fig. 2. Example images (top) and the corresponding annotations (bottom) collected from (a) CQ500, (b) SinoCT and (c) PDDCA.

early stopping configured but use a hard threshold to stop the training. In our testing, the threshold is set as 50 epochs for the two models. For the rest, all are set to have a patience of 10 epochs for fine-tuning. One thing that needs to be mentioned is that there are two models in the YOLOv8 + SAM method and they are trained separately and an additional bounding box from the segmentation mask needs to be generated for the YOLOv8 training. For the U-Mamba model, there are two structures proposed in the original paper, one uses SSM block in only the bottleneck block (U-Mamba Bot) and the other in all the encoder blocks (U-Mamba Enc). Both of the two models are tested in our experiments. On top of the model outputs, a post-processing step is added to remove extra foreground regions in the image and only keep the two largest ones. This step is to ensure the results are less affected by the artifacts or noises in the image.

Evaluation Metrics. The results of the models are evaluated based on three metrics: the Dice Similarity Coefficient (DSC), Mean Surface Distance (MSD), and the Hausdorff Distance (HD). The DSC gives a general estimation of the segmentation accuracy by calculating the overlap between the ground truth set A and predicted set B divided by their summation:

$$DSC = \frac{2|A \cap B|}{|A| + |B|} \quad (1)$$

The MSD and HD evaluate the accuracy based on the boundary distance between the ground truth surface S_{GT} and the predicted surface S_P. The MSD gives an overall surface accuracy evaluation by averaging the minimum distances between points from

Table 2. Summary of the models in training.

Model	Input Size	Optimization	Batch Size	Training Epoch
DeepLabv3 + [25]	512 × 512	SGD	5	31
U^2-Net [13]	320 × 320	Adam	5	18
nnU-Net [14]	512 × 512	SGD	16	50
YOLOv8 + SAM [27]	1024 × 1024	Adam	8	55
U-Mamba Bot [26]	512 × 512	SGD	16	50
U-Mamba Enc [26]	512 × 512	SGD	16	50

one surface to the other:

$$MSD(S_{GT}, S_P) = \frac{1}{|S_{GT}| + |S_P|}\left(\sum_{x \in S_{GT}} d(x, S_P) + \sum_{y \in S_P} d(y, S_{GT})\right) \quad (2)$$

while the HD focuses on the worst case by finding the maximum distance between the two surfaces:

$$HD(S_{GT}, S_P) = \max\left\{\sup_{x \in S_{GT}} d(x,\ S_P) \sup_{y \in S_P} d(y,\ S_{GT})\right\} \quad (3)$$

In the above formulas, the distance calculation $d(\cdot,\ \cdot)$ is based on the Euclidean distance and the actual distance in mm is used by multiplying the pixel distance with the pixel spacing in the evaluation.

3 Experiment Results

The testing results of different models on the CQ500 + SinoCT dataset are shown in Table 3. It can be observed that compared with the traditional CNN-based models, the transformer and mamba based models have better performance in terms of the dice scores. However, for the boundary distance indices, it shows that the U^2-Net achieved the best performance in terms of both the averaged mean distance (MSD) and the maximum distance values (HD). This can be because the U^2-Net processes the data in multi-scale and is capable of detecting the fine details of the image. The result images of one case inferred using different models are shown in Fig. 3.

Although the patient-level data splitting is adopted in the training and testing for the CQ500 + SinoCT dataset to avoid the data leakage issue. The results are still not enough to show how the models generalize on new data collected from other machines in different environment. This is very critical in the medical field applications for obtaining consistent accurate results regardless of the equipment setup differences. Therefore, to test on the generalization capability, the models trained on the CQ500 + SinoCT dataset are directly used for the inference of CGH and PDDCA dataset.

The CGH data is quite similar to the CQ500 and SinoCT data in terms of the pixel spacing. The inference results are shown in Table 4. It can be seen that the performance

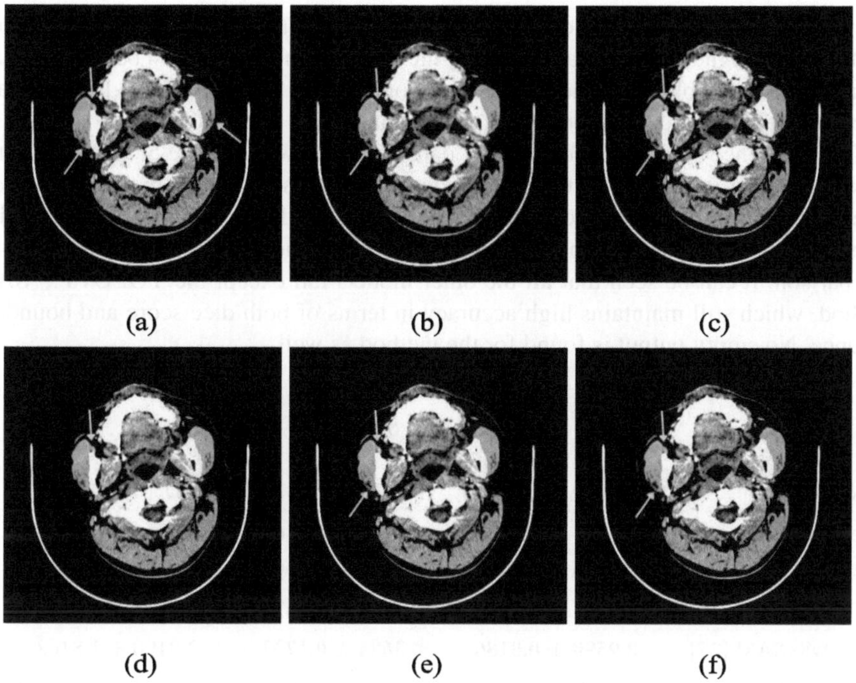

Fig. 3. Example segmentation result of (a) DeepLabv3 +, (b) U^2-Net, (c) nnUNet, (d) YOLOv8 + SAM, (e) U-Mamba Bot and (f) U-Mamba Enc. The boundary is the ground truth contour while the shaded area is the segmentation result. The missegmented regions are pointed out with arrows.

Table 3. Inference results of the CQ500 + SinoCT dataset.

Model	DSC ↑	MSD ↓	HD ↓
DeepLabv3 + [25]	0.9524 ± 0.0494	0.4832 ± 1.3029	3.2278 ± 3.8852
U^2-Net [13]	0.9632 ± 0.0434	**0.1503 ± 0.3693**	**1.0638 ± 1.4763**
nnU-Net [14]	0.9756 ± 0.0152	0.2013 ± 0.1192	2.1748 ± 1.3449
YOLOv8 + SAM [27]	0.9680 ± 0.0132	0.2758 ± 0.1086	1.7213 ± 1.3429
U-Mamba Bot [26]	**0.9771 ± 0.0129**	0.1936 ± 0.1072	2.1116 ± 1.3630
U-Mamba Enc [26]	0.9738 ± 0.0153	0.2166 ± 0.1226	2.3386 ± 1.4528

drops for all the models. Compared with other methods, the performance degradation of YOLOv8 + SAM is the smallest. This is because both of the YOLO and SAM models are pre-trained on very large datasets with extensive of data augmentations. In addition, the hierarchical process takes advantage of the complementary strengths of these two architectures: the multiscale accurate object detection and pixel-perfect segmentation, making it less sensitive to the data variations. In contrast, although the boundary accuracy of the U^2-Net is not bad, the dice score drops dramatically. For the mamba-based models, they have similar performance drop to the CNN models. This can be because the mamba models are designed to excel at capturing the long range dependency but not the data variation in this problem. The addition globle content even makes the boundary accuracy much worse by mis-segment other part of the images.

The inference results of the PDDCA dataset are shown in Table 5. Because of the large difference in pixel spacing settings during the data capturing process, there are cases of no foreground detected. Those cases are excluded in the indices calculation to avoid extreme values and the amount is recorded in the last column of the table for comparison. It can be seen that all the other models fail except the YOLOv8 + SAM method, which still maintains high accuracy in terms of both dice score and boundary distance. No empty output is found for the method as well.

Table 4. Inference results of the CGH dataset.

Model	DSC ↑	MSD ↓	HD ↓
DeepLabv3+ [25]	0.8767 ± 0.0691	1.0118 ± 0.7461	7.0614 ± 4.0155
U^2-Net [13]	0.8956 ± 0.0850	0.4209 ± 0.3996	**2.6887 ± 1.7461**
nnU-Net [14]	0.9197 ± 0.0959	0.8479 ± 1.1439	6.0524 ± 5.2612
YOLOv8+SAM [27]	**0.9590 ± 0.0186**	**0.3684 ± 0.1732**	2.9190 ± 1.8168
U-Mamba Bot [26]	0.8567 ± 0.1669	4.4190 ± 12.0009	16.5474 ± 25.7116
U-Mamba Enc [26]	0.8891 ± 0.1285	3.0326 ± 7.8912	11.5702 ± 18.1007

Table 5. Inference results of the PDDCA dataset. $|E|$ is the number of cases that no foreground pixel is detected.

Model	DSC ↑	MSD ↓	HD ↓	$\|E\|$ ↓
DeepLabv3 + [25]	0.2847 ± 0.2192	2.8344 ± 0.8535	88.1303 ± 45.7425	16
U^2-Net [13]	0.5509 ± 0.1566	4.0840 ± 5.6850	34.0889 ± 13.6650	2
nnU-Net [14]	0.4909 ± 0.3543	8.9158 ± 19.9029	36.4273 ± 44.3102	4
YOLOv8 + SAM [27]	**0.9264 ± 0.0647**	**0.7123 ± 0.8242**	**3.2807 ± 3.2133**	**0**
U-Mamba Bot [26]	0.4421 ± 0.3407	14.7329 ± 25.2667	54.2920 ± 52.4215	2
U-Mamba Enc [26]	0.5803 ± 0.2870	6.9416 ± 11.5894	44.9855 ± 43.5871	1

4 Discussion and Conclusion

In this paper, we introduce a novel dataset specifically designed for masseter muscle (MM) segmentation, collected from diverse sources. To the best of our knowledge, this is the first dataset developed for this purpose. Based on the data collected, a series of experiments is carried out to evaluate different AI models for MM segmentation. The performance is evaluated based on different indices including dice scores and surface distances. The generalization capability of the models on unseen datasets captured from different environments is assessed as well.

The results show that testing on the same dataset of training, the U-Mamba models can have better results in terms of dice score and the U^2-Net performs best in terms of the boundary distance metrics. However, when testing on datasets that have never been used in the training, the performance degrades differently for the models evaluated. The two-stage model YOLOv8 + SAM has shown superior performance even when dealing with the dataset that has large difference from the training set. This is likely due to the multi-scale object handling capabilities of both YOLOv8 and SAM.

Due to limited manpower, the proposed data set is collected from only 4 sources with 1000 + images, more comprehensive studies can be done by collecting extra data from different sources. The data preparation process needs a good amount of manual work to extract and label the data. It will be good if an automatic image detection and preparation pipeline is designed for generating the larger dataset. Future studies could extend the scope to include 3D data, which can give more information about the region of interest for helping diagnosis. For the model part, further studies can be carried out based on the findings of the current results, for example, further improvements based on the combination of different models. The pipeline and methods can also be applied to the other region detection problems in the CT image.

In summary, this study represents a significant step forward in the field of MM segmentation by presenting a new dataset, developing an evaluation pipeline, and providing a detailed analysis of various AI models. The findings offer valuable insights and a foundation for future research, including dataset expansion, automated processing, and methodological advancements.

Acknowledgments. The authors extend their gratitude to Mr. Bernard Koh Tiang Guan and Mr. Narayan Venkataraman from Data Management & Informatics, CGH, for their dedicated assistance in data preparation. The authors acknowledge all the authors of the employed datasets and models for making their valuable resources publicly available.

References

1. Gaudy, J.-F., Zouaoui, A., Bravetti, P., Charrier, J.-L., Guettaf, A.: Functional organization of the human masseter muscle. Surg. Radiol. Anat. **22**, 181–190 (2000)
2. Pan, Y., Chen, S., Shen, L., Pei, Y., Zhang, Y., Xu, T.: Thickness change of masseter muscles and the surrounding soft tissues in female patients during orthodontic treatment: a retrospective study. BMC Oral Health **20**, 1–10 (2020)

3. de Smit, M.J., Nijholt, W., Bakker, M.H., Visser, A.: The predictive value of masticatory function for adverse health outcomes in older adults: a systematic review. J. Nutrit. Health Aging **28**(5), 100210 (2024)
4. Eo, P.S., et al.: The effect of orthognathic surgery on changes in masticatory muscle volume in patients with facial asymmetry. J. Craniofac. Surg. **33**(5), 1288–1293 (2022)
5. Qin, H., Pei, Y., Guo, Y., Ma, G., Xu, T., Zha, H.: Masseter segmentation from computed tomography using feature-enhanced nested residual neural network. In: International Workshop on Machine Learning in Medical Imaging, pp. 355–362. Springer (2018)
6. Ng, H.P., Ong, S.H., Foong, K.W.C., Goh, P.-S., Nowinski, W.L.: Masseter segmentation using an improved watershed algorithm with unsupervised classification. Comput. Biol. Med. **38**(2), 171–184 (2008)
7. Ng, H.P., et al.: 3D segmentation and quantification of a masticatory muscle from MR data using patient-specific models and matching distributions. J. Digit. Imag. **22**, 449–462 (2009)
8. Guan, Q., Zhang, B., Long, H., Hu, H., Zhuang, X., Hu, Y.: A modified distance regularized level set evolution for masseter segmentation. In 2016 8th International Conference on Information Technology in Medicine and Education (ITME), pp. 16–19. IEEE (2016)
9. Chen, W., et al.: Deep learning vs. atlas-based models for fast auto-segmentation of the masticatory muscles on head and neck CT images. Rad. Oncol. **15**, 1–10 (2020)
10. Santoro, M., et al.: Recent applications of artificial intelligence in radiotherapy: where we are and beyond. Appl. Sci. **12**(7), 3223 (2022)
11. Orhan, K., Yazici, G., Kolsuz, M.E., Kafa, N., Bayrakdar, I.S., C¸ elik, O¨.: An artificial intelligence hypothetical approach for masseter muscle segmentation on ultrasonography in patients with bruxism. J. Adv. Oral Res. **12**(2), 206–213 (2021)
12. Ronneberger, O., Fischer, P., Brox, T.: U-net: convolutional networks for biomedical image segmentation. In: Medical Image Computing and Computer Assisted Intervention–MICCAI 2015, pp. 234–241. Springer (2015)
13. Qin, X., Zhang, Z., Huang, C., Dehghan, M., Zaiane, O.R., Jagersand, M.: U2-Net: Going deeper with nested U-structure for salient object detection. Pattern Recogn. **106**, 107404 (2020)
14. Isensee, F., Jaeger, P.F., Kohl, S.A.A., Petersen, J., Maier-Hein, K.H.: nnU-Net: a self-configuring method for deep learning-based biomedical image segmentation. Nat. Methods **18**(2), 203–211 (2021)
15. C¸ ic¸ek, O¨ ., Abdulkadir, A., Lienkamp, S.S., Brox, T., Ronneberger, O.: 3D U-Net: learning dense volumetric segmentation from sparse annotation. In Medical Image Computing and Computer-Assisted Intervention–MICCAI 2016, pp. 424– 432. Springer (2016)
16. Jiang, Y., et al.: Automatic masseter muscle accurate segmentation from CBCT using deep learning-based model. J. Clin. Med. **12**(1), 55 (2022)
17. Keser, G., Bayrakdar, I. S., Pekiner, F. N., C¸ elik, O¨ ., Orhan, K.: A deep learning approach for masseter muscle segmentation on ultrasonography. J. Ultrasonogr. **22**(91), e204–e208 (2022)
18. Zhang, Y., et al.: Masseter muscle segmentation from Cone-Beam CT images using generative adversarial network. In 2019 IEEE 16th International Symposium on Biomedical Imaging (ISBI 2019), pp. 1188–1192. IEEE (2019)
19. Pan, Y., Wang, Y., Li, G., Chen, S., Xu, T.: Validity and reliability of masseter muscles segmentation from the transverse sections of Cone-Beam CT scans compared with MRI scans. Int. J. Comput. Assist. Radiol. Surg., 1–9 (2022)
20. Iyer, A., et al.: Prospectively-validated deep learning model for segmenting swallowing and chewing structures in CT. Phys. Med. Biol. **67**(2), 024001 (2022)
21. Zhong, Y., Pei, Y., Nie, K., Zhang, Y., Xu, T., Zha, H.: Bi-graph reasoning for masticatory muscle segmentation from cone-beam computed tomography. IEEE Trans. Med. Imag. (2023)

22. Chilamkurthy, S., et al.: Deep learning algorithms for tection of critical findings in head CT scans: a retrospective study. Lancet **392**(10162), 2388–2396 (2018)
23. Raudaschl, P.F., et al.: Evaluation of segmentation methods on head and neck CT: auto-segmentation challenge 2015. Med. Phys. **44**(5), 2020–2036 (2017)
24. Hooper, S.M., et al.: Impact of upstream medical image processing on downstream performance of a head CT triage neural network. Radiol.: Artif. Intell. **3**(4), e200229 (2021)
25. Chen, L.-C., Zhu, Y., Papandreou, G., Schroff, F., Adam, H.: Encoderdecoder with Atrous separable convolution for semantic image segmentation. In Proceedings of the European Conference on Computer Vision (ECCV), pp. 801–818 (2018)
26. Ma, J., Li, F., Wang, B.: U-mamba: enhancing long-range dependency for biomedical image segmentation (2024). arXiv preprint: arXiv:2401.04722
27. Pandey, S., Chen, K.-F., Dam, E.B.: Comprehensive multimodal segmentation in medical imaging: Combining YOLOv8 with SAM and HQ-SAM models. In Proceedings of the IEEE/CVF International Conference on Computer Vision, pp. 2592–2598 (2023)
28. Ma, J., He, Y., Li, F., Han, L., You, C., Wang, B.: Segment anything in medical images. Nat. Commun. **15**, 654 (2024)
29. Ma, J.: SOTA-MedSeg: state-of-the-art medical image segmentation (2024). https://github.com/JunMa11/SOTA-MedSeg. Accessed 28 Nov 2024

DiffNovo: A Transformer-Diffusion Model for De Novo Peptide Sequencing

Shiva Ebrahimi, Jiancheng Li, and Xuan Guo(✉)

Department of Computer Science and Engineering, University of North Texas, Denton, TX 76207, USA
{Shiva.Ebrahimi,Jiancheng.Li,Xuan.Guo}@unt.edu

Abstract. Accurate identification of peptides from Data-Independent Acquisition (DIA) mass spectrometry data is crucial for discovering novel peptides and proteins. In this study, we introduce DiffNovo, which integrates transformer-based architectures with a diffusion model to enhance de novo peptide sequencing for DIA data. The experimental results demonstrate that DiffNovo outperforms existing state-of-the-art models in recall and precision for de novo peptide identification. DiffNovo achieves an average improvement of 20.7% in amino acid precision and an increase of 57.2% in peptide precision across various datasets. These substantial gains highlight the efficacy of DiffNovo in delivering accurate peptide sequencing from DIA mass spectrometry data.

Keywords: de novo peptide sequencing · data-independent acquisition (DIA) · transformer · diffusion

1 Introduction

Mass spectrometry has emerged as a pivotal technology for the comprehensive analysis of protein content in biological samples, playing a central role in advancing proteomics by enabling the identification of unknown peptides [1–4]. Among various approaches, Data-Independent Acquisition (DIA) has gained prominence for its ability to perform comprehensive and unbiased analysis of complex samples. Unlike traditional Data-Dependent Acquisition (DDA) methods, DIA simultaneously fragments all ions within specified mass ranges, offering broad coverage and high sensitivity [5, 6]. This makes it particularly suited for detecting low-abundance peptides that might otherwise be overlooked. Despite these advantages, the inherent complexity of DIA data presents significant challenges for de novo peptide sequencing that require sophisticated computational approaches. Although considerable progress has been made in peptide identification methods, handling highly multiplexed spectra, often containing multiple precursor ions, remains a formidable obstacle [5, 6]. Consequently, there is a growing need for solutions that enhance both sensitivity and accuracy in de novo peptide identification.

This work was supported by the National Center for Complementary & Integrative Health and the National Institute of General Medical Sciences of the National Institutes of Health [R01AT011618].

T. Aldwairi et al. (Eds.): BICOB 2025, CCIS 2535, pp. 96–104, 2025.
https://doi.org/10.1007/978-3-031-94039-2_8

Recent advances in deep learning have significantly enhanced de novo peptide sequencing by enabling models to better navigate the complexities of mass spectrometry data [3, 7, 8]. However, while DDA has seen extensive model development, DIA presents unique challenges—such as multiplexed signals, overlapping fragment ions, and higher noise levels—that have only recently begun to be addressed with specialized deep-learning approaches. For instance, DeepNovoDIA [9] extended the original DeepNovo framework to process DIA MS/MS spectra, yet it suffered from low peptide-level accuracy and relied heavily on feature extraction tools to preprocess the data. Building on these initial efforts, PepNet [10] introduced a fully convolutional neural network that seamlessly integrates both DIA and DDA data, marking a step toward more versatile sequencing methods. Similarly, BiATNovo [11] was designed for dual compatibility with DDA and DIA data by incorporating a CNN-based spectrum encoder and dual attention-based decoders. This dual approach, employing both independent and synchronous decoding strategies, enhances the precision of peptide sequencing by effectively capturing the bidirectional nature of the sequencing process. More recently, Transformer-DIA [12] leveraged transformer architectures to improve performance on DIA data, though it still required a preliminary feature extraction step. Cascadia [13] was developed as a refinement of the Casanova framework specifically tailored for DIA spectra, achieving further performance enhancements through architectural optimizations. Despite these promising developments, deep learning models for DIA data analysis remain less mature compared to those for DDA.

In this study, we present DiffNovo—a novel model that extends the Transformer-DIA framework by integrating a transformer-based architecture with a discrete diffusion model. This innovative combination leverages the transformer's sequence-learning strengths alongside the diffusion model's probabilistic framework to effectively manage noisy and highly multiplexed DIA mass spectrometry data, resulting in more accurate identification of complex and novel peptides.

Our experiments on three DIA datasets show that DiffNovo significantly outperforms current state-of-the-art models in both accuracy and sensitivity, thereby substantially improving peptide identification. By enhancing precision and robustness, DiffNovo has the potential to accelerate the discovery of novel peptides and proteins, with important implications for drug discovery, biomarker development, and disease research. The remainder of this paper is organized as follows: the Methods section details DiffNovo's architecture, including the integration of the transformer-based model with a discrete diffusion model and the elimination of high-dimensional fragment ion computations during preprocessing, and the Experimental Results section presents our comparative analyses. DiffNovo is freely available under the GNU GPL license at https://github.com/Biocomputing-Research-Group/DiffNovo.

2 Methods

In this section, we present a novel approach to de novo peptide sequencing that integrates a transformer-based model with a diffusion model, combining their complementary strengths within a unified framework. Diffusion models, a class of generative models, operate by iteratively perturbing data into noise and reconstructing it through a learned reverse process, enabling precise sequence prediction through step-by-step refinement. When applied to de novo peptide sequencing, this iterative process enhances the reconstruction of peptide sequences with increased confidence. The transformer component effectively learns from the highly multiplexed MS/MS spectra generated by DIA, leveraging its capability to identify peptide sequences from the most relevant peaks. Meanwhile, the diffusion model progressively refines these predictions, addressing the inherent challenges of noise and complexity in DIA data. Together, this cohesive framework achieves significant improvements in sensitivity and accuracy, offering a robust solution for the identification of precise peptide sequences (Fig. 1).

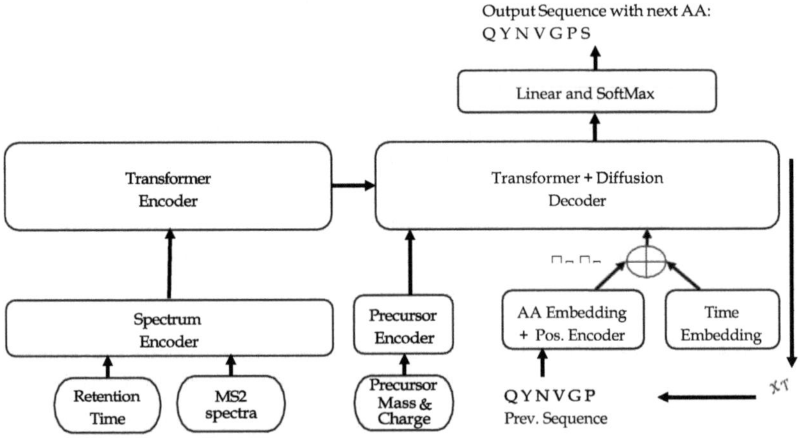

Fig. 1. DiffNovo Architecture. DiffNovo integrates transformer and diffusion models into a unified framework. The spectrum encoder generates latent spectral representations, while the decoder predicts and refines the next amino acid at each decoding step. In the forward process, $Q_t x_t$ computes the probability distribution of possible next states, where x_t represents the predicted peptide sequence at timestep t, and Q_t is the transition matrix.

2.1 Transformer-Based De Novo Peptide Sequencing

The success of transformer-based de novo peptide sequencing models lies in their effective utilization of self-attention and cross-attention mechanisms. Selfattention enables the model to capture pairwise dependencies among fragment ions in MS/MS spectra and amino acids in peptide sequences, fostering a deep contextual understanding. Multihead attention enables the model to simultaneously focus on different aspects of the input sequence, allowing it to capture complex patterns in fragment ions and peptide sequences more effectively. The cross-attention mechanism aligns mass spectrum embeddings with precursor ion embeddings, prioritizing the most relevant elements of both sequences. This architecture facilitates nuanced interpretation of MS/MS data, leading to accurate peptide sequence inference.

2.2 Diffusion Models for Generative Peptide Sequencing

Diffusion models complement transformers by enhancing generative capabilities, allowing the reconstruction of peptide sequences not present in existing databases. By introducing gradual corruption and denoising processes, diffusion models can identify novel peptide sequences and post-translational modifications (PTMs) not captured by traditional database search methods. Inspired by recent developments in discrete state-space diffusion models [14–16], our approach employs a transition matrix to iteratively add and remove noise, refining peptide sequence predictions through multiple timesteps.

2.3 DiffNovo: A Unified Framework

DiffNovo integrates transformer and diffusion models within a cohesive framework, consisting of an encoder and a decoder as an extension of TransformerDIA. The spectrum encoder processes MS/MS spectra, retention times, and precursor information, utilizing self-attention layers to generate a latent representation of the spectra.

Spectrum Encoder: As illustrated in Fig. 2, the spectrum encoder processes MS/MS spectra, where each spectrum comprises (m/z, intensity) values along with retention times. Positional embeddings are applied to these vectors, followed by concatenation and processing through multi-layer perceptrons (MLPs). This output is fed into the transformer encoder, which applies self-attention and feedforward layers for spectral representation. The self-attention mechanism, a key component of transformer architecture, allows the model to effectively learn the relationships between peaks by dynamically weighing the importance of different input elements.

Decoder Architecture: The decoder takes three inputs: the spectrum representation vector, previously predicted peptide sequences, and precursor information. Precursor data, embedded using sinusoidal functions, is summed to form precursor embeddings. Self-attention layers with sinusoidal position embeddings capture relationships among amino acids in the sequence. A cross-attention layer aligns peptide sequences with the encoded spectrum, translating spectra into peptide sequences via a final softmax layer.

3 Forward and Reverse Processes in Diffusion Models

The forward process gradually adds noise to peptide sequences, starting from an initial state as an uncorrupted peptide sequence and proceeding through t timesteps using a transition matrix Q_t. We define a transition matrix $Q_t = \mathcal{N}(\sqrt{\alpha_t}x_{t-1}, (1 - \alpha_t)I)$, where $\alpha_t = 1 - \beta_t$. The noise level at each timestep t is controlled by a cosine scheduler, with β_t defined as $\beta_t = 1 - \cos\left(\frac{t\pi}{2T}\right)^2$. Using this scheduler, the sequences x_t are gradually corrupted with increasing noise over time. The reverse process removes this noise step-by-step, reconstructing the original sequence x_0 from the corrupted x_T. The reverse model is parameterized by neural network weights θ, predicting the probabilities of each possible state for x_{t-1}. The reverse process utilizes a Transformer decoder instead of relying on transposed transition matrices. During the training phase of a diffusion model, the model learns to reverse the diffusion process by predicting the previous state based on

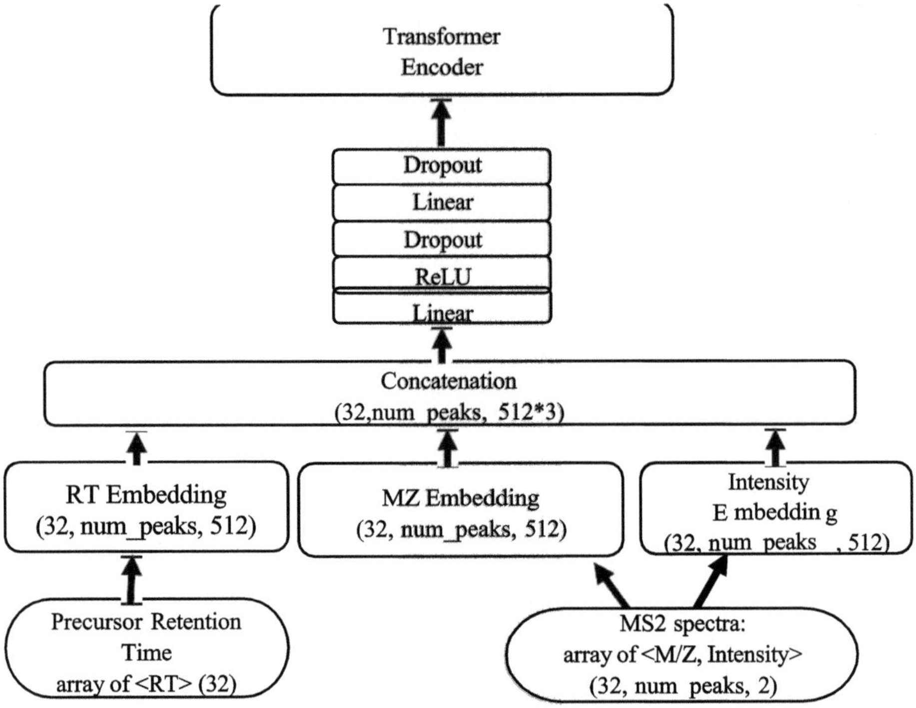

Fig. 2. Spectrum Encoder. The spectrum encoder processes MS2 spectra, including m/z, intensity, and retention time, by generating sinusoidal embeddings for each and encoding them into three distinct vectors. These vectors are then concatenated and passed through multi-layer perceptrons (MLPs). The resulting output is fed into the transformer encoder, which utilizes self-attention and feedforward layers to generate a comprehensive spectral representation.

the current noisy state. The Transformer decoder is particularly effective for predicting previous states by attending to the current noisy predicted sequences while leveraging learned representations of the sequence structure.

Model training and Loss Function. Training is supervised using a crossentropy loss function that compares the reconstructed peptide sequences generated by the model with the ground truth peptides:

$$L_{reverse} = -\sum x_0 \log p\theta \left(x_0|x_t\right),$$

where x_0 represents the ground truth sequence, and $p_\theta(x_0\ x_t)$ is the predicted probability.

DiffNovo employs a transformer architecture with nine layers in both the encoder and decoder, an embedding size of 512, and eight attention heads, resulting in approximately 50 million trainable parameters. The feedforward dimension is set to 1024. Training is performed using the top 150 most intense peaks per spectrum, with a batch size of 32 spectra and a weight decay of 10^{-5}. Beam search is applied with a beam size of 5 during training and 20 during evaluation. The learning rate follows a warm-up strategy, increasing linearly from 0 to 1.0 x 10^{-4} over 100,000 steps, before transitioning to a cosine decay schedule. Models are trained for 110 epochs on six GeForce RTX 2080 GPUs, with an average training time of 37 h.

4 Decoding with Beam Search and Post-correction

The decoding process involves three steps: beam search, denoising, and postcorrection. Beam search predicts amino acid sequences iteratively, selecting the top candidates at each step. The denoising algorithm refines predictions iteratively, improving the sequence accuracy over multiple steps. Finally, the postcorrection algorithm ensures alignment with the experimental precursor mass, systematically examining and substituting amino acids to achieve a closer mass match, ensuring precise peptide sequence identification.

5 Experiments and Results

5.1 Datasets

To evaluate the performance of the proposed DiffNovo model, we utilized datasets released along with the DeepNovo-DIA model. These datasets were sourced from the MassIVE repository (accession number MSV000082368) and include three datasets from *Homo sapiens*: urinary tract infection (UTI), ovarian cysts (OC), and plasma. Spectrum annotation was performed by selecting the five spectra with the highest intensity values for each precursor ion, as these are considered the most informative for peptide sequence identification. The annotated datasets contained 33,710 spectra for plasma, 98,146 spectra for UTI, and 76,094 spectra for OC. Each dataset was randomly divided into training, validation, and testing subsets in a ratio of 70%, 15%, and 15%, respectively. To prevent data leakage, spectra corresponding to the same feature—as defined in the original DeepNovo-DIA feature files—were kept together during the data split. Note that all benchmarked methods were trained on the same annotated files, with additional feature files provided for DeepNovo-DIA and Cascadia. In addition, Cascadia was fine-tuned using three DIA datasets from a model pre-trained on DDA data, as described in the original study [13].

5.2 Evaluation Metrics

We used three evaluation metrics commonly employed by state-of-the-art models:

- **Amino Acid Precision:** The proportion of correctly matched amino acids among all predicted amino acids.
- **Amino Acid Recall:** The proportion of correctly matched amino acids relative to the total amino acids in the ground truth peptide sequences.
- **Peptide Precision:** The proportion of fully accurate predicted peptides among all predicted peptide sequences.

A prediction was deemed correct if the mass difference between the predicted and ground truth amino acids was below 0.1 Da, with prefix or suffix mass deviations under 0.5 Da.

5.3 Results

To evaluate the performance of DiffNovo, we compared its amino acid recall, amino acid precision, and peptide precision with three state-of-the-art (SOTA) methods: DeepNovo-DIA, PepNet, and Cascadia. The results across three DIA datasets—UTI, OC, and Plasma—are summarized in Tables 1 and 2.

Amino Acid Precision and Recall. Table 1 shows that DiffNovo consistently outperforms existing methods (DeepNovo-DIA, PepNet, and Cascadia) on both amino acid recall and precision for the UTI and OC datasets, highlighting the robustness of our approach. Specifically, for the UTI dataset, DiffNovo's recall (0.665) is about 17% higher than the next best method (0.566), and its precision (0.675) is roughly 10% higher than the next best method (0.612). For the OC dataset, DiffNovo's recall (0.749) exceeds the next best approach (0.547) by about 37%, while also surpassing the best precision (0.743) by approximately 23%. On the Plasma dataset, although DiffNovo does not surpass the top-performing methods in every metric, it still achieves competitive results, with a recall (0.663) that is higher than both PepNet (0.509) and a precision (0.675) that outperforms PepNet (0.491). These findings suggest that DiffNovo provides a strong overall performance advantage on most DIA datasets, although further optimization may be needed for more challenging datasets like Plasma.

Peptide Precision. Table 2 presents the peptide-level precision results on the UTI, OC, and Plasma datasets. DiffNovo achieves the highest precision on the UTI dataset (0.659), which is about 63% higher than the second-best method (0.403). Similarly, on the OC dataset, DiffNovo attains a precision of 0.735, representing a 75% improvement over the second-best method (0.417). For the Plasma dataset, although DiffNovo (0.632) surpasses DeepNovo-DIA (0.351) by 80%, it is still below PepNet (0.550) and Cascadia (0.736). Overall, these findings indicate that DiffNovo consistently delivers strong performance across multiple DIA datasets, while additional optimization may be necessary to improve results on more challenging datasets such as Plasma.

Table 1. Comparison of DiffNovo and other state-of-the-art (SOTA) methods on amino acid recall and precision across three DIA datasets.

DIA Datasets	Amino Acid Recall DiffNovo DeepNovo-DIA PepNet Cascadia				Amino Acid Precision DiffNovo DeepNovo-DIAPepNet			Cascadia
UTI	**0.665**	0.566	0.396	0.293	**0.675**	0.612	0.391	0.341
OC	**0.749**	0.547	0.455	0.286	**0.743**	0.601	0.451	0.349
Plasma	0.663	0.689	0.509	**0.704**	0.675	0.702	0.491 / 0.725*	**0.782**

The best entry is in bold. * Indicates the positional accuracy reported in [10].

Table 2. Comparison of DiffNovo and other state-of-the-art (SOTA) methods on peptide precision across three DIA datasets.

DIA Datasets	DiffNovo	DeepNovo-DIA	Peptide Precision PepNet	Cascadia
UTI	**0.659**	0.403	0.296	0.291
OC	**0.735**	0.417	0.368	0.301
Plasma	0.632	0.351	0.550/0.530*/0.664+	**0.736**

The best entry is in bold. + Indicates the filtered peptide-level accuracy, and * indicates the peptide-level accuracy reported in [10].

6 Conclusion

DiffNovo integrates a transformer architecture with diffusion mechanisms for de novo peptide sequencing, offering significant performance gains over existing state-of-the-art methods. By eliminating the computation of high-dimensional product ion peptides in the preprocessing step, DiffNovo reduces both time complexity and GPU usage. Its enhanced spectrum encoder further improves performance at both the amino acid and peptide levels across multiple DIA datasets. In particular, DiffNovo consistently outperforms DeepNovo-DIA, PepNet, and Cascadia on the UTI and OC datasets, achieving substantial increases in peptide precision. Although Cascadia achieves the highest scores on the Plasma dataset, DiffNovo remains competitive, underscoring its robustness and generalizability. Overall, DiffNovo's ability to enhance sequencing accuracy while maintaining computational efficiency highlights its potential to advance proteomics research and broaden the application of deep learning in mass spectrometry-based peptide identification. Future work could involve refining model architectures and training strategies to optimize performance across diverse datasets.

References

1. Xuan, G., et al.: Sipros ensemble improves database searching and filtering for complex metaproteomics. Bioinformatics **34**(5), 795–802 (2018)
2. Xuan Guo, Y., Meng, N.Y., Pan, Y.: Cloud computing for detecting high-order genome-wide epistatic interaction via dynamic clustering. BMC Bioinform. **15**, 1–16 (2014)
3. Feng, S., Sterzenbach, R., Guo, X.: Deep learning for peptide identification from metaproteomics datasets. J. Proteom. **247**, p. 104316 (2021)
4. Li, J., Xiong, Y., Feng, S., Pan, C., Guo, X.: CloudProteoAnalyzer: scalable processing of big data from proteomics using cloud computing. Bioinform. Adv. **4**(1), vbae024 (2024)
5. Doerr, A.: DIA mass spectrometry. Nat. Methods **12**(1), 35 (2015)
6. Zhang, F., Ge, W., Ruan, G., Cai, X., Guo, T.: Data-independent acquisition mass spectrometry-based proteomics and software tools: a glimpse in 2020. Proteomics **20**(17–18), 1900276 (2020)
7. Feng, S., et al.: MetaLP: An integrative linear programming method for protein inference in metaproteomics. PLOS Comput. Biol. **18**(10), e1010603 (2022)
8. Ebrahimi, S., Guo, X.: Deep active learning for de novo peptide sequencing from data-independent-acquisition mass spectrometry. Tech. rep. EasyChair (2022)

9. Tran, N.H., et al.: Deep learning enables de novo peptide sequencing from data-independent-acquisition mass spectrometry. Nat. Methods **16**(1), 63–66 (2019)
10. Liu, K., Ye, Y., Li, S., Tang, H.: Accurate de novo peptide sequencing using fully convolutional neural networks. Nat. Commun. **14**(1), 7974 (2023)
11. Wu, S., Luan, Z., Fu, Z., Wang, Q., Guo, T.: BiATNovo: a self-attention based bidirectional peptide sequencing method. In: bioRxiv, pp. 2023–05 (2023)
12. Ebrahimi, S., Guo, X.: Transformer-based de novo peptide sequencing for data-independent acquisition mass spectrometry. In: 2023 IEEE 23rd International Conference on Bioinformatics and Bioengineering (BIBE), pp. 28–35. IEEE (2023)
13. Sanders, J., et al.: A transformer model for de novo sequencing of data-independent acquisition mass spectrometry data. In: bioRxiv, pp. 2024–06 (2024)
14. Sohl-Dickstein, J., Weiss, E., Maheswaranathan, N., Ganguli, S.: Deep unsupervised learning using nonequilibrium thermodynamics. In: International Conference on Machine Learning, pp. 2256–2265. PMLR (2015)
15. Hoogeboom, E., Nielsen, D., Jaini, P., Forré, P., Welling, M.: Argmax flows and multinomial diffusion: Learning categorical distributions. In: Advances in Neural Information Processing Systems, vol. 34, pp. 12454–12465 (2021)
16. Austin, J., Johnson, D.D., Ho, J., Tarlow, D., Van Den Berg, R.: Structured denoising diffusion models in discrete statespaces. In: Advances in Neural Information Processing Systems, vol. 34, pp. 17981–17993 (2021)

Disease Networks for Disease Clustering Using Graph Convolution Networks

Hisham Al-Mubaid[1] and Tamer Aldwairi[2(✉)]

[1] University of Houston, Clear Lake, Houston, USA
hisham@uhcl.edu
[2] Temple University, Philadelphia, USA
aldwairi@temple.edu

Abstract. This paper presents a technique for disease clustering using the gene ontology biological process annotation terms along with disease gene association networks with deep learning embedding. In this paper, we explore and leverage the application of Graph Convolution Networks *GCN* in bioinformatics; specifically in disease clustering and disease relationships. GCNs allow for aggregating two sources of information on the disease by integrating the disease network structure with disease biological process (*bp*) annotations from the gene ontology. The paper serves as an exploration study to examine the feasibility of integrating network structures with feature vectors for disease clustering. We obtained the disease relationship network using disease-shared genes from *OMIM* and *DisGeNET* and the disease *bp* features from the Gene Ontology annotation *GOA* database. The clustering method was performed with the embedding feature space and the experimental results show improved clustering outcomes with the feature embeddings using *GCN*.

1 Introduction

Disease similarity is an important task in bioinformatics and other biomedical science areas like disease biology, disease interaction, and comorbidity [1–3]. Disease similarity and relationship studies and methods contain multiple problems including finding similar diseases, disease relationships, and disease clustering [1–5]. Moreover, disease similarity and clustering work involve various types of data and multiple information sources, for example, disease-associated genes, disease symptoms, disease phenotypes, proteins, therapeutic chemicals, and more [1, 4, 17–19]. There are important applications that benefit from the outcomes of disease clustering and similarity research like drug repurposing, disease prediction, and understanding disease mechanisms, which can lead to improving treatment plans and patient outcomes [18].

Typically, integrating more than one source of information and data type can yield better results in disease similarity and clustering [6–9]. For example, if we combine the disease network (as *a graph*) with disease *process* annotation information in the vector space model we should obtain improved results. As one source of information, a disease network represents each disease i as a vertex v_i in the network graph and an edge e_i represents a relationship, *i.e., a link*, between disease i and disease j (Fig. 1). On the

T. Aldwairi et al. (Eds.): BICOB 2025, CCIS 2535, pp. 105–114, 2025.
https://doi.org/10.1007/978-3-031-94039-2_9

other hand, process annotation information of a disease *i* can be cast as a feature vector *xi* in a feature space model with dimensionality *d*. Graph convolution networks (*GCN*) are one type of graph neural networks that allow us to aggregate graph structure information with disease process annotation information viewed as feature vectors [2, 5, 6]. In this paper, we would like to explore and leverage the use of *GCN* in disease clustering and disease relationships. The presented methodology allows for integrating two sources of information on diseases using *GCN*: (1) the disease network structure, and (2) disease biological process (*bp*) annotations from the *Gene Ontology (GO)*. Therefore, this paper presents an exploration study to examine the feasibility of integrating network structures with feature vectors for disease clustering using *graph convolution networks* [1–6, 9, 10, 15, 16].

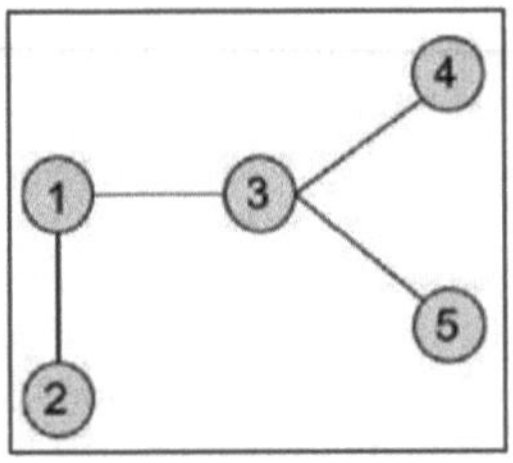

Fig. 1. A sample disease network consisting of five nodes where each node *vi* represents one disease, and each edge is a relationship between two diseases.

2 Literature Review

The integration of gene ontology (GO) annotations with disease clustering represents a critical advancement in understanding the biological processes underlying various diseases. Gene Ontology, a comprehensive framework for annotating genes with biological terms, provides insights into gene function, interaction, and pathways. GO terms are organized into three major categories: biological process (BP), molecular function (MF), and cellular component (CC). The biological process (BP) category, in particular, has proven crucial for clustering diseases based on shared molecular mechanisms and pathways [12]. Recent studies have leveraged these annotations to classify diseases by their molecular profiles, revealing new insights into disease mechanisms and potential therapeutic strategies.

Gene Ontology terms have been extensively used to link genes with their associated biological processes, enabling researchers to explore the functional relationships between genes and diseases. A key challenge in computational biology is accurately clustering diseases based on genetic and functional similarities. Several efforts have explored the integration of Gene Ontology (GO) into machine learning algorithms for gene expression analysis. For instance, [28] presents a GO-based feature selection approach, embedding ontology information into machine learning models for more accurate gene classification. This method incorporates grouping information from GO terms such as cellular component, molecular function, and biological process. The authors demonstrated the effectiveness of their approach on eight gene expression datasets, revealing its ability to identify significant ontology terms for classification tasks. The tool can assist geneticists in identifying affected genes in transcriptomic data, aiding in diagnosis, prognosis, and patient treatment planning. Building on these methodologies, [24] improved the interpretation of *Alzheimer's* disease gene expression data by curating *GO* annotations focused on neuroinflammation. Their findings highlighted novel biological processes, such as gliogenesis, and demonstrated the enhanced applicability of GO for understanding AD's molecular mechanisms, particularly in neuroinflammation. Similarly, [25] introduced *RepTB*, a Gene Ontology-based drug repurposing framework for tuberculosis (TB), which utilizes molecular function correlations to predict novel drug-target interactions. By creating a network of drug and target nodes enriched with GO terms, RepTB identifies promising drug candidates for TB, prioritizing essential M. tuberculosis targets. The approach was validated by identifying potential targets such as FolP1, Tmk, Dut, and MenB, with further exploration of new chemical space for testing. This method offers a scalable model for drug repurposing, applicable to TB and other infectious diseases.

In another effort, Sanjak et. al. (2024) [26] utilized an ontology-enriched knowledge graph to cluster rare diseases based on shared etiology and pathophysiology. Using data from 3242 rare diseases, enriched with gene and phenotype ontologies, biological pathways, and small molecule activity data, the researchers constructed a knowledge graph and applied node embedding techniques for clustering. Their approach produced 37 disease clusters, validated through semantic similarity and feature enrichment analyses, revealing related genes and potential links to drug targets. This framework demonstrates a scalable method for clustering diseases and supports further exploration in predictive modeling and drug repurposing efforts.

Furthermore, [27] applied unsupervised machine learning to predict drugs for lifestyle- related diseases based on symptom features. Among the findings, the drug clusters generated by the Bisecting K-means algorithm demonstrated strong similarity in gene ontology annotations of drug targets, alongside chemical ontology and structural features. This approach highlights the role of gene ontology in ensuring biologically meaningful clustering. A web application, SDLDpred, was developed to provide clinicians with decision-making support for early LSD treatment.

Moreover, [29] developed GeNetOntology, a tool that leverages Gene Ontology (GO) for feature selection in gene expression data analysis. Unlike traditional computational methods, this approach integrates biological domain knowledge into the gene selection process by embedding GO terms into a machine learning algorithm. The tool employs a Grouping, Scoring, and Modeling (G-S-M) approach to identify significant GO terms, which are then used for two-class classification tasks. Tested on 11 gene expression datasets, the method successfully identifies disease-related ontology terms and has potential applications in disease understanding and diagnosis. Finally, [30] explored the use of functional annotations of disease genes, specifically the biological process aspect from the Gene Ontology, for disease clustering. The authors apply an unsupervised machine learning approach to group diseases based on the semantic similarity of their associated gene functions. Their method successfully clusters diseases by leveraging Gene Ontology annotations, and the results are validated using both MeSH and DO ontologies. The research highlights the reliability of using functional gene annotations for clustering diseases, which can aid in understanding disease mechanisms, gene-disease associations, and potential drug repurposing opportunities.

3 Materials and Methods

Typically, GCN_s leverage the concept of local neighborhoods to exploit more sources of information for a given data point. In general, graph machine learning, including graph neural networks and graph convolution networks can be applied to either node classification or link prediction [7]. The neighborhood of a given node consists of the nodes that are connected by an edge with the given node; e.g., in Fig. 1, nodes 2 and 3 comprise the neighborhood of node 1. The feature representation (*e.g.*, feature vector) of each node is aggregated (*e.g.*, averaged) with the (feature) representation of its neighborhood nodes. Here, in this work, we would like to use *GCN* to integrate the *disease-bp* associations (*i.e., disease biological processes annotations*) with the disease network in a disease clustering (and relationship) task. The disease network is constructed as a graph with vertices (diseases) and edges (associated shared genes). Each disease *i* is represented as a vertex *vi* in the disease network graph, and an edge *eij* represents a relationship (or *link*) between disease *i* and disease *j*. Therefore, the disease network is a graph *G*. In other words, we construct the disease network as a graph *G* using the disease-associated genes based on the shared, common genes among the diseases. Given a graph $G = (V, E)$ where $V = \{v_i\}_{i=1}^{N}$ and $E = \{e_{i,j}\}_{i,j=1}^{N}$, e_{ij}=1 if there is an edge between v_i and v_j; and 0 otherwise. An edge between e_{ij} between vertex v_i and v_j indicates that diseases *i* and *j* have at least one shared gene (one gene in common as disease-associated gene for both disease *i* and disease *j*). Let us construct from *G* an adjacency matrix *A* (binary values 0 and 1; and $A \in R^{N \times N}$). Let a feature matrix $X = \{x_i\}_{i=1}^{N}$ where each disease *i* has a feature vector $x_i \in R^d$; and *d* is the feature space dimensionality, *N* is the number of diseases (vertices), the feature matrix is $X \in R^{N \times d}$. So, we use *GCN* to obtain node (disease) embeddings that capture the graph structures and the feature values. Then, a clustering method (*e.g., k-means, DBSCAN, spectral clustering*) is applied with the latent embeddings to obtain clustered data. In the neural network model, *GCN*, we use a nonlinear activation function $f(\cdot)$ with each input $H^{(l)}$ of layer l such that $H^{(l+1)} = f(H^{(l)}, A)$, and *A* is the adjacency matrix.

So, initially: let $H^{(0)}$=X and at each layer:

$$H^{(l)} = \sigma(\tilde{D}^{-1/2}\tilde{A}\tilde{D}^{-1/2}H^{(l-1)}W^{(l)}) \tag{1}$$

{this is the GCN model, with layer-wise propagation rule}
where

$$\tilde{A} = A + I_N \tag{2}$$

{I_N is the identity matrix, here just adding self-connections to the graph}

$\tilde{D}$ is diagonal node degree matrix of $\tilde{A}$ such that

$$\tilde{D}_{ii} = \sum\nolimits_j \tilde{A}_{ij} \tag{3}$$

$W^{(1)} \in R^{cxd}$, *weight matrix in layer l,*
and

$$\sigma(x) = ReLU(x) = max(0, x) \tag{4}$$

$$\begin{aligned} \{for\ example:\ H^{(0)} &= X, \\ H^{(1)} &= ReLU(\tilde{D}^{-1/2}\tilde{A}\tilde{D}^{-1/2}\,H^{(0)}W^{(1)}), \\ H^{(2)} &= ReLU(\tilde{D}^{-1/2}\tilde{A}\tilde{D}^{-1/2}\,H^{(1)}W^{(2)}), \\ &\ldots and\ so\ on.\} \end{aligned} \tag{5}$$

Figure 2 illustrates the architecture of the proposed GCN model. We designed the proposed model with three layers (as shown in Fig. 2) where the first two layers use the main *GGGGGG* layer-wise propagation rule in Eq. (1). The third layer, $H^{(3)}$, is a linear projection representing the embeddings of disease feature data in an h3-dimensional space:

$$H^{(3)} = H^{(2)}W_3 \tag{6}$$

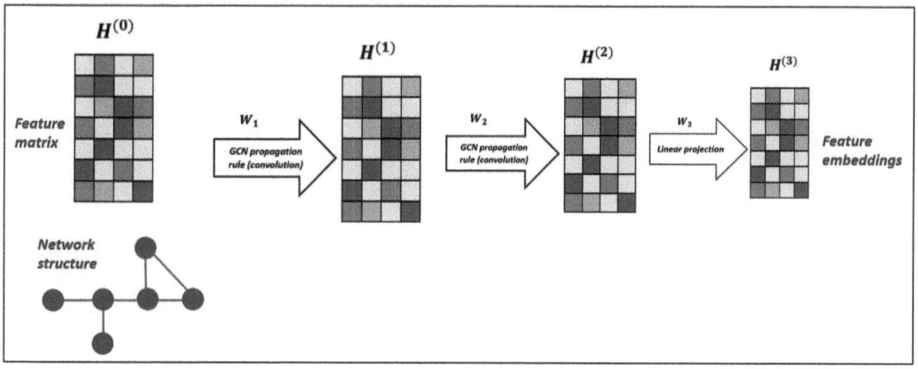

Fig. 2. Schematic view of the proposed *GCN* architecture.

4 Results

We obtained the disease relationship network using disease-shared genes from *OMIM* and *DisGeNET* [13, 21] and the disease *bp* features from the *Gene Ontology* annotation *GOA* database [22]. The clustering process was performed with the *k-means* algorithm on the full *bp* feature set initially, then with the *GCN* embedding feature space. Brady et. al. (2024) [20] use the path length in the disease ontology (*DO*) as a measure of disease similarity and closeness, which is adapted to validate clustering results [20, 23]. For evaluation and experiments, we collected three datasets, as shown in Table 1. The first dataset *D1a40* consists of 40 diseases with 220 *bp* features in the input feature space, and the output embedding dimensionality is 40. The other two datasets, *D2a50* and *D3a60,* details are in Table 1. After clustering $\{k = 2\}$ with the *k-means* clustering algorithm is done, we compare the quality of clustering by finding the shortest path length (in the *DO* [20]) of diseases within one cluster (*intra-cluster distance*) versus diseases in two clusters (*inter-cluster distance*). Therefore, we used the *DO* as a validation method to examine the relationships between diseases within one cluster and across two clusters. For example, in Fig. 3 (a), the two diseases d_1–d_2 should have a shorter distance, within *DO* ontology hierarchy, compared with diseases $d1$–$d3$.

Table 1. The preliminary experimental results of clustering using *GCN* embeddings of disease process features for disease clustering

Dataset	BP feature matrix	Embeddings feature matrix	Clustering with BP features		Clustering with embeddings	
			intra-cluster	inter-cluster	intra-cluster	inter-cluster
D1a40	40 × 220	40 × 60	3.15	4.05	2.95	4.40
D2a50	50 × 220	50 × 70	3.10	3.95	3.05	4.30
D3a60	60 × 220	60 × 75	2.85	3.95	2.70	4.55

Table 2. The difference in the DO distances between intra-vs inter-cluster pairs in clustering with full bp features and with embeddings.

Datasets	Difference: intraCluster – interCluster	
	Clustering with BP features	Clustering with embeddings
D1a40	0.90	1.45
D2a50	0.85	1.25
D3a60	1.10	1.85

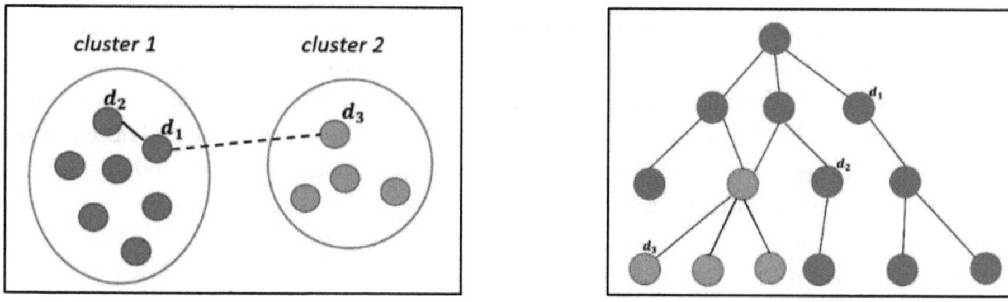

(a) (b)

Fig. 3. (a) Disease clusters: The disease pair (d_1, d_2) which consists of the two diseases d_1 and d_2 belong to the same cluster while disease pair (d_1, d_3) is dispersed in two clusters. (b) Disease Ontology (*DO*) is used here to calculate and obtain the *DO* distance (*shortest path length*) between $d_1 - d_2$ versus $dd_1 - d_3$ in the *DO* hierarchy.

5 Discussion

In disease clustering work, we would like to find more similar diseases based on the information and data sources used in clustering [4, 9, 20, 23]. This clustering procedure is an unsupervised machine learning process that attempts to group data points, diseases in this case, into two (or more) distinct groups, called clusters, based on certain patterns found in the data. So, disease clustering can be viewed as a similarity or relationship process since two diseases assigned to the same cluster should be more similar compared to two diseases assigned to two different clusters; Figure 3 (a). The quality of input data affects the clustering results. In this work, we employ *GCN* to prepare the input data for clustering. The *GCN* aggregates two data modalities on diseases. In the evaluation, we examined the shortest path length in the *DO* hierarchy between two diseases $d1$and $d2$ in the disease pair ($d1$, $d2$). A disease pair (d_i , d_i) can be either *intra-cluster* where both diseases d_i , *di* are in the same cluster, or *inter-cluster* pair where *di* and *di* are assigned to two distinct clusters. From the clustered data, we randomly collected *inter-cluster* and *intra-cluster* disease pairs. Then we obtained the *DO* distance (shortest path length) in the *DO* hierarchy between the two diseases in each disease pair. We performed the clustering step with the full *bp* disease features and with the *GCN* deep embedding as shown in Table 1. That is, we performed disease clustering with the full disease *bp* features and with the aggregated feature space by *GCN*. The results in Table 1 include the average *DO* distance for *intra-cluster* and *inter-cluster* disease pairs with the full *bp* feature set and with *CGN* embeddings. As shown in Table 2, the difference in *DO* distance between *inter-* vs *intra-cluster* pairs is higher (1.45) with *GCN* compared with full bp features (0.90) in the first experiment *D1a40*. In the second experiment, *D2a50*, the difference is 0.85 with the raw data but higher (better) 1.25 with *GCN* features. Also, the third experiment shows a clear clustering improvement with *CGN* (1.85) compared with raw *bp* data (1.10). These results indicate that the diseases placed in one cluster are more similar, according to the *DO* hierarchy, to one another when we use the *CGN* embedding disease features.

Results explained: in Table 1, the value 3.15 represents the average path length of all *intra-cluster* disease pairs in experiment *D1a40* (when clustering with the full *bp* features); whereas 4.05 is the average path length of all inter-cluster disease pairs in that same experiment. The difference $\{4.05 - 3.15 = 0.90\}$ is shown in Table 2. These results in Table 1 and Table 2 show clear improvements in the clustering results when *GCN* is used compared with the full *bp* feature set.

6 Conclusions

We presented a methodology and study for disease relationship and clustering using graph convolution neural networks by integrating disease networks with disease biological process features. Disease biological process features are obtained from the *biological process bp* taxonomy in the Gene Ontology. This work serves as proof of concept and the experimental results are limited but encouraging. Clearly, the evaluation presented in this paper for the proposed methodology is not enough but can serve as a first step and a proof of concept for combining two data modalities. Disease semantic relationships are infused in the graph and network structure while the disease *bp* features comprise and convey the disease functional information. The results show that after using a graph convolution network for aggregating disease information the clustering results are improved based on the disease semantic distance in the Disease Ontology.

References

1. Chen, Y., et al.: CoGO: a contrastive learning framework to predict disease similarity based on gene network and ontology structure (2022)
2. Kipf, T.N., Welling, M.: Semi-supervised classification with graph convolutional networks (2017)
3. Li, Y., et al.: Evaluating disease similarity based on gene network reconstruction and representation. Bioinformatics **37**, 3579–3587 (2021)
4. Sanjak, J., Zhu, Q., Mathe, E.A.: Clustering rare diseases within an ontology-enriched knowledge graph. J. Am. Med. Inform. Assoc. (2023). PMID: 36824742; PMCID: PMC9949046
5. Parisot, S., et al.: Disease prediction using graph convolutional networks: application to Autism Spectrum Disorder and Alzheimer's disease. Med. Image Anal. **48**, 117–130 (2018)
6. Li, S., Zhang, R.: A novel interactive deep cascade spectral graph convolutional network with multi-relational graphs for disease prediction. Neural Netw. **175** (2024)
7. Lu, H., Uddin, S.: Disease prediction using graph machine learning based on electronic health data: a review of approaches and trends. Healthcare (Basel). **11**(7), 1031 (2023). PMID: 37046958
8. Oss Boll, H., et al.: Graph neural networks for clinical risk prediction based on electronic health records: a survey. J. Biomed. Inform. **151** (2024)
9. Al Jreidy, M., et al.: Clustering using graph convolution networks. Prog. Artif. Intell. (2024). https://doi.org/10.1007/s13748-023-00310-z
10. Kipf, T.N., Welling, M.: Semi-supervised classification with graph convolutional networks. arXiv preprint: arXiv:1609.02907 (2016)

11. Kim, C.Y., et al.: HumanNet v3: an improved database of human gene networks for disease research. Nucleic Acids Res. **50**, D632–D639 (2022). https://doi.org/10.1093/nar/gkab1048
12. Ashburner, M., et al.: Gene ontology: tool for the unification of biology. Nat. Genet. **25**, 25–29 (2000)
13. Pinero, J., et al.: The DisGeNET knowledge platform for disease genomics: 2019 update. Nucleic Acids Res. **48**, D845–D855 (2020)
14. Yuan, Y., Bar-Joseph, Z.: GCNG: graph convolutional networks for inferring gene interaction from spatial transcriptomics data. Genome Biol. **21**, 300 (2020). https://doi.org/10.1186/s13059-020-02214-w
15. Li, Q., Han, Z., Wu, X.-M.: Deeper insights into graph convolutional networks for semi-supervised learning. In: The Thirty-Second AAAI Conference on Artificial Intelligence (AAAI-18)
16. del Valle, E.P.G., et al.: Disease networks and their contribution to disease understanding: a review of their evolution, techniques and data sources. J. Biomed. Inform. **94**, 103206 (2019). https://doi.org/10.1016/j.jbi.2019.103206
17. Sánchez-Valle, J., Valencia, A.: Molecular bases of comorbidities: present and future perspectives. Trends in Genet. **39**(10) (2023). ISSN 0168–9525, https://doi.org/10.1016/j.tig.2023.06.003
18. del Valle, E.P.G., et al.: DisMaNET: a network-based tool to cross map disease vocabularies. Comput. Methods Programs Biomed. **207** (2021). https://doi.org/10.1016/j.cmpb.2021.106233
19. Brady, D., Al-Mubaid, H.: Disease clustering with process annotations from gene ontology. In: Proceedings of 16th International Conference on Bioinformatics and Computational Biology (BICOB-2024), EPiC Series in Computing, vol.101 (2024)
20. Amberger, J.S., Bocchini, C.A., Scott, A.F., Hamosh, A.: OMIM.org: leveraging knowledge across phenotype–gene relationships. Nucleic Acids Res. **47**(D1), D1038–D1043 (2019). https://doi.org/10.1093/nar/gky1151
21. Aleksander, S.A., et al.: The gene ontology knowledgebase in 2023. Genetics **224**(1), iyad031 (2023). https://doi.org/10.1093/genetics/iyad031
22. Brady, D., Al-Mubaid, H.: Disease similarity and disease clustering. In: Proceedings of Southwestern Data Science Conference SDSC-2024 (2024)
23. Kramarz, B., et al.: Gene ontology curation of neuroinflammation biology improves the interpretation of Alzheimer's disease gene expression data. J. Alzheimer's Dis. **75**(4), 1417–1435 (2020). https://doi.org/10.3233/JAD-200207
24. Passi, A., et al.: RepTB: a gene ontology based drug repurposing approach for tuberculosis. J. Cheminform. **10**, 24 (2018). https://doi.org/10.1186/s13321-018-0276-9
25. Sanjak, J., et al.: Clustering rare diseases within an ontology-enriched knowledge graph. J. Am. Med. Inform. Assoc. **31**(1), 154–164 (2024)
26. Bhattacharjee, S., Saha, B., Saha, S. Symptom-based drug prediction of lifestyle-related chronic diseases using unsupervised machine learning techniques. Comput. Biol. Med. **174**, 108413 (2024). https://doi.org/10.1016/j.compbiomed.2024.108413
27. Yousef, M., Sayıcı, A., Bakir-Gungor, B.: Integrating gene ontology based grouping and ranking into the machine learning algorithm for gene expression data analysis. In: Kotsis, G., et al. (eds.) Database and Expert Systems Applications - DEXA 2021 Workshops. DEXA 2021. Communications in Computer and Information Science, vol. 1479. Springer, Cham (2021). https://doi.org/10.1007/978-3-030-87101-7_20

28. Ersoz, N.S., Bakir-Gungor, B., Yousef, M.: GeNetOntology: identifying affected gene ontology terms via grouping, scoring, and modeling of gene expression data utilizing biological knowledge-based machine learning. Front. Genet. **14**, 1139082 (2023). https://doi.org/10.3389/fgene.2023.1139082
29. Al-Mubaid, H., Aldwairi, T.: Utilizing functional annotation of disease genes for disease clustering. In: Proceedings of International Conference on Bioinformatics and Computing Biology (BICOB-2023). EPiC Series in Computing, vol. 92, 2023, pp. 58–71 (2023). https://easychair.org/publications/paper/gVMg

Parallel Implementation of the Stochastic Reaction Diffusion Master Equation

Md Mustafijur Rahman[1](✉), Keisha Cook[2], and Roger B. Sidje[3]

[1] Department of Computational, Engineering, and Mathematical Sciences (CEMS), Texas A&M University - San Antonio, San Antonio, TX, USA
mrahman24@crimson.ua.edu

[2] School of Mathematical and Statistical Sciences, Clemson University, Clemson, SC, USA
keisha@clemson.edu

[3] Department of Mathematics, The University of Alabama, Tuscaloosa, AL, USA
roger.b.sidje@ua.edu

Abstract. The Reaction Diffusion Master Equation (RDME) provides a stochastic framework for modeling spatially distributed biochemical systems. RDME is commonly analyzed using numerical simulations of trajectories over time, employing Monte Carlo methods, notably the Stochastic Simulation Algorithm (SSA), which incorporates potential randomness. However, the SSA is computationally demanding, and this grows worse when applied to largescale systems. In the present work, we use a parallel implementation to speed up RDME simulations. We leverage OpenMP with FORTRAN to effectively distribute calculations across several processors. This allows us to markedly decrease runtime while preserving accuracy, facilitating the simulation of time-consuming systems with the use of high-performance computing.

Keywords: reaction diffusion master equation · stochastic simulation algorithm · parallel computing · OpenMP · chemical master equation

1 Introduction

Parallel computing plays a pivotal role in advancing computationally demanding tasks, and in systems biology it can thus facilitate the efficient analysis of complex biological systems. Biological systems at the cellular level entail random interactions among species that have small copy numbers. The chemical master equation (CME) [10] is a mathematical framework used to describe the stochastic behavior of chemical reactions in a well-mixed system. It provides a way to model the time evolution of the probability distribution of the number of molecules of each species in a system, considering both the discrete characteristics of molecules and the stochastic nature of their interactions.

The stochastic simulation algorithm (SSA) [8, 9] is a computational method employed to obtain sample realizations guided by the chemical kinetics that underpin the CME. The SSA is a powerful tool for simulating complex biochemical networks, as it can take into account the stochasticity of the system and can generate statistically significant outcomes. We refer to [11] for a background overview. Although beneficial, the

T. Aldwairi et al. (Eds.): BICOB 2025, CCIS 2535, pp. 115–126, 2025.
https://doi.org/10.1007/978-3-031-94039-2_10

SSA algorithm is time-consuming since it is based on realizations of the states over time. As a result, many variants and computational strategies have been proposed, including even using grid computing [3].

In contrast to conventional chemical reaction models that assume that reactions occur instantaneously and uniformly throughout the system, cellular environments are not uniform in space, and spatial heterogeneity is a common feature. Experimental data from multiple sources [5, 6, 13], demonstrate the significance of considering both the stochastic characteristics and spatial distribution of a system to explain its behavior. The premise is that molecules in stochastic reaction-diffusion systems function as points that engage in continuous Brownian motion in space. Whenever two molecules come within a defined reaction radius, bimolecular chemical reactions occur instantly. This modeling approach is referred to as spatially continuous diffusion-limited reaction (SCDLR), which was originally proposed by Smoluchowski [15]. In contrast to these methods, when we treat the diffusion at a molecular level as a special set of reactions in the CME, we arrive at the reaction-diffusion master equation (RDME) [7].

RDME is a mathematical framework utilized to characterize the dynamics of stochastic chemical reaction systems that incorporate diffusion processes. The equation describes the time-dependent probability distribution function (PDF) of the system state, similar to the CME, but with a significantly higher dimensionality. In RDME, the space is divided into compartments. Each compartment is regarded as well-mixed, and reactions within a specific compartment are assumed to be consistent with the homogeneous case. Also, molecules can get diffusive transfers between adjacent compartments. The state of the system at any given time is defined by the number of molecules of each species in each compartment. Mathematically, the RDME is the forward Kolmogorov equation for a continuous-time jump Markov process. One viable approach to address the RDME is to produce samples from it and gather statistics using a Monte Carlo method. The works in [2, 16] illustrate some examples of the SSA applied to reaction-diffusion systems in one space dimension.

In this paper, we utilize the RDME framework in two biological models, integrating reaction and diffusion processes, and simulate stochastic trajectories via the SSA algorithm. Due to the substantial computational expense of executing the SSA within the RDME framework, we utilize parallel computing to minimize simulation duration. These findings highlight the significance of parallel computing in enhancing large-scale biological simulations, prompting additional investigation into sophisticated computational methodologies.

The rest of the paper is organized as follows: In Sect. 2, we summarize the CME, SSA, and RDME. Section 3 discusses parallel implementation techniques. Sections 4 and 5 provide numerical test results, discussions and conclusion.

2 Methods

2.1 CME

Consider a biochemical system of N species $S_1,\dots, S_N$, represented by the state vector $\mathbf{X}(t) = [X_1(t),\dots, X_N(t)]^T$, where each $X_i(t)$ is a non-negative integer denoting the number of molecules of species S_i at time t. The species are interacting via M reaction channels, denoted by $R_1,\dots, R_M$. Transitions between states occur only when a reaction occurs. Each reaction R_j entails a state change (stoichiometric) vector $\boldsymbol{v}_j = [v_{1j},\ldots, v_{Nj}]^T$ that represents the change when the reaction occurs. The components of the state change vector are integers displaying the increase or decrease of the number of copies of each species after the associated reaction occurs. If the system is in state $\mathbf{X}(t) = \mathbf{x}$, and reaction j occurs, then the system transitions to state $\mathbf{x} + \boldsymbol{v}_j$. Each state $\mathbf{x}$ entails M propensities $a_1(\mathbf{x}),\ldots, a_M(\mathbf{x})$ that determine the relative chance of each reaction occurring if the system is in state $\mathbf{x}$. The propensities are defined by the requirement that, given $\mathbf{X}(t) = \mathbf{x}$, $a_j(\mathbf{x})dt$ is the probability of reaction j occurring in the next infinitesimal time interval $[t, t + dt]$, where the dependence on time has now been made explicit.

This system can be represented as a Markov process, with its dynamics governed by the Chemical Master Equation (CME) [10]

$$\frac{dP(\mathrm{x}, \mathrm{t})}{dt} = \sum_{j=1}^{M} [a_j(\mathrm{x} - v_j)P(\mathrm{x} - \mathrm{v_j}, \mathrm{t}) - \mathrm{a_j}(\mathrm{x})\mathrm{P}(\mathrm{x}, \mathrm{t})], \tag{1}$$

where the function $P(\mathbf{x}, t)$ denotes the probability that $\mathbf{X}(t)$ will be $\mathbf{x}$. The CME (1) can be expressed in an equivalent matrix–vector form by enumerating all the states. If there are n possible states $\mathbf{x}_1,\ldots, \mathbf{x}_n$, the CME is represented as a system of linear ordinary differential equations (ODEs),

$$\dot{\mathbf{P}}(t) = \boldsymbol{M} \cdot \mathrm{P}(t), \tag{2}$$

where $\boldsymbol{M}$ is the transition matrix that describes the chemical reactions and the probability vector $\mathbf{P} = [p_1,\ldots, p_n]^T$ is such that each component $p_i = P(\mathbf{x}_i, t) = Prob(\mathbf{x}(t) = \mathbf{x}_i)$, the probability of being at state $\mathbf{x}_i$ at time t, for $i = 1,\ldots, n$.

2.2 SSA

The SSA is extensively employed to determine the trajectories of chemical reaction networks described by the CME, particularly because of the "curse of dimensionality", that makes finding an analytical solution impractical. As the number of species and reactions escalates, the possible states of the system grow exponentially, making it computationally infeasible to monitor all states directly. SSA addresses this by providing a numerical solution that simulates the time evolution of the system via discrete events. SSA operates by simulating individual reactions one by one, calculating the propensity functions for each possible reaction, and employing random numbers to determine the time and order of reactions.

As outlined in Algorithm 1, the SSA uses random numbers to draw sample paths of the state vector $\mathbf{X}(t)$ over time. The algorithm determines the time τ at which the next reaction will occur and which reaction j will occur next. Each reaction has a propensity function a_j, which captures how likely that one reaction will occur in $[t, t + \tau)$ with rate $a_j(\mathbf{X}(t))\tau$. After τ and j are determined, the state of the system is updated by $\mathbf{X}(t + \tau) = \mathbf{X}(t) + \boldsymbol{v}_j$ until the final time t_f. The SSA ensures that each time step τ is small enough so that only one reaction occurs. Algorithm 1 gives the pseudocode for the SSA [9].

Algorithm 1 SSA

Initial State: $\mathbf{X}(t_0) = x_0$, Final Time: t_f
$t := t_0$;
while $t < t_f$ **do**
  $x = \mathbf{X}(t)$
  Evaluate $\alpha_j(x)$ and $\alpha_0(x)$;
  Generate two random numbers r_1, r_2 from $U(0,1)$
  $j := min\{j, \sum_{j'=1}^{j} \alpha_{j'}(x) > r_2\alpha_0(x)\}$;
  $\tau := ln(1/r_1)/\alpha_0(x)$;
  $\tau := min\{\tau, t_f - t\}$;
  $t := t + \tau$;
  Check for negative values
  $\mathbf{X}(t+\tau) := x + \boldsymbol{\nu}_j$
end while

By generating a large number of stochastic trajectories, the SSA method can provide an accurate approximation of the behavior of the reaction network over time, even in cases where an analytical solution is not feasible due to the curse of dimensionality.

2.3 RDME

Assume the domain Ω is divided into compartments (voxels). We designate the compartments with V_k, $k = 1,\ldots, K$. Molecules in each compartment can interact with one another within that compartment and can also diffuse across boundaries to adjacent compartments. Both the reaction and diffusion processes are regarded as random processes. Let, $X_{i,k}(t)$ be the number of molecules of species S_i within compartment V_k at time t. Then each species in the domain is represented by the subvector $\mathbf{X}_i(t) = [X_{i,1}(t),\ldots, X_{i,K}(t)]$, $i = 1,\ldots, N$. That means $\mathbf{X}_1$ denotes the subvector which represents the species 1 in all compartments, $\mathbf{X}_2$ represents species 2, and so on. Consequently, the state vector of the system is $\mathbf{X} = [\mathbf{X}_1,\ldots, \mathbf{X}_N]$. The diffusion propensity function $d_{i,j,k}$ and the state change vector $\boldsymbol{\mu}_{k,j}$ characterize the dynamics of the diffusion of species S_i from compartment V_k to V_j. The vector $\boldsymbol{\mu}_{k,j}$ possesses a length of K with -1 in the kth position, 1 in the jth position, and 0 elsewhere. Given, $\mathbf{X}(t) = \mathbf{x}$ the diffusion master equation (DME) can be expressed by

$$\frac{dP(\mathrm{x,t})}{dt} = \sum_{i=1}^{N}\sum_{k=1}^{K}\sum_{j=1}^{K} [d_{i,j,k}(\mathrm{x}_i - \mu_{k,j})P(\mathrm{x}_1, \ldots, \mathrm{x}_i - \mu_{k,j}, \ldots \mathrm{x}_N, t) - d_{i,j,k}(\mathrm{x}_i)P(\mathrm{x}, t)]. \tag{3}$$

The diffusion propensity function $d_{i,j,k}(\mathbf{x}_i) = D/l^2$, for $k = j \pm 1$, where D is the diffusion rate and l is the characteristic length of the compartment. If $\boldsymbol{D}$ represents the transition matrix characterizing the diffusion of molecules, the corresponding matrix-vector form can be written by

$$\dot{\mathbf{P}}(t) = D \cdot \mathbf{P}(t). \tag{4}$$

Combining equations (2) and (4) we obtain the matrix-vector form of the RDME,

$$\dot{\mathbf{P}}(t) = \boldsymbol{M} \cdot \mathbf{P}(t) + \boldsymbol{D} \cdot \mathbf{P}(t). \tag{5}$$

Equation (5) constitutes a system of linear constant coefficient ODEs and gives us more possible states than the CME, and so its corresponding $\boldsymbol{M}$ is substantially extended to represent species in compartments. Overall, the RDME captures both the reaction and diffusion, hence it has a much higher dimensionality.

3 Parallel Implementation

3.1 Parallel Processing

HPC systems often rely on parallel processing, where multiple processors or cores work simultaneously on different parts of a task. We can call the original algorithm a task. This task is then split into different parts among multiple CPU cores. The different parts are running simultaneously. While it is possible to use as many cores as a platform permits, doing so comes with an overhead in the communications. Given our experimental context, we will be using a small number of cores to divide our algorithm. This is usually 1, 2, 4, 8, or 16 cores. There are two main paradigms for organizing parallel processes. One of them, called shared memory, occurs when different processors need to access the same memory location. The other paradigm is distributed memory, which occurs when each processor has its own private memory location.

3.2 OpenMP

Our numerical tests use OpenMP interfacing with the Fortran programming language. "Open" means that the standard application is accessible to anyone and "MP" stands for multi-processing. Within OpenMP, there are two types of parallelism. Coarse-grain parallelism, where the target domain is distributed into a set of subdomains that are distributed over different processors and fine-grain parallelism, where the work of do-loops is distributed over different processors where each processor has a set of iterations. Our focus is on fine-grain parallelism. In order to activate OpenMP codes, the directive sentinel !$OMP is employed. The exclamation point "!" is used so that a normal compiler will interpret the line as a comment. An OpenMP compiler will read the line as a directive and enable the parallel capabilities. To define a parallel region in OpenMP, we employ the following directive pair.

!$OMP PARALLEL DO

enter code here

!$OMP END PARALLEL DO

The code in between the two directives is executed in parallel, while the serial regions before and after the parallel directive pair are executed sequentially. Each core or thread is assigned a specific set of computations. The main thread, called the master thread, waits for all of the threads inside a parallel region to finish before continuing the sequential code outside of the parallel region. We refer to this process as synchronization. Figure 1 demonstrates a serial program divided into three threads, which are executed in parallel.

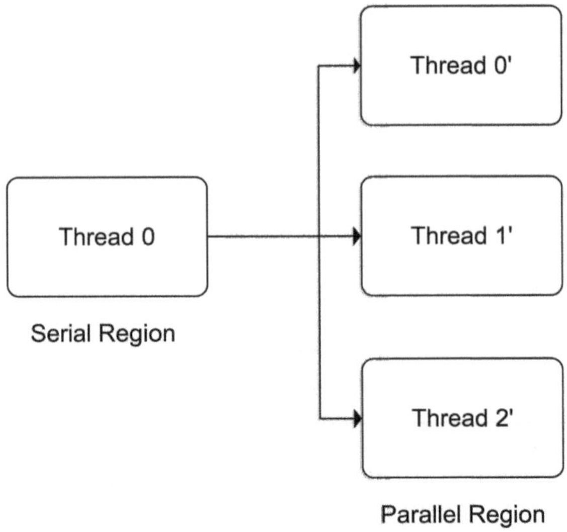

Fig. 1. Diagram of parallel implementation into three threads.

3.3 Speed-Up and Parallel Efficiency

When measuring the effects of parallel implementation, we focus on two terms; speed-up and parallel efficiency. We use the following formulas [12]:

$$\text{Speed} - \text{Up} = \frac{time_1}{time_p} \qquad \text{Parallel Efficiency} = \frac{\frac{time_1}{time_p}}{p}$$

where $time_p$ is the computational time using p cores and $p = 1, 2, 4, 8, 16$.

4 Numerical Results

4.1 Metapopulation Model

Metapopulation models, utilized to analyze populations segmented into subpopulations linked by migration, often employ reaction-diffusion models to describe spatial dynamics [4]. In systems biology, these models provide a framework for studying the emergent behaviors of complex biological networks. In a metapopulation model using the RDME, each subpopulation's population is modeled as a stochastic process characterized by random occurrences of infection, and migration events. The diffusion term captures the movement of individuals between adjacent subpopulations, while the reaction term describes infection within each subpopulation. We consider a basic reaction scheme

$$\beta \xrightarrow{c_1} \alpha, \tag{6}$$

$$\alpha + \beta \xrightarrow{C_2} 2\beta. \tag{7}$$

In the context of our metapopulation model, α particles represent normal particles (healthy individuals) and β particles represent active particles (infected individuals) intermingling from area to area. This framework is sometimes referred to as the SIS model [1]. We consider two neighboring areas (i.e. compartments) where both the healthy and infected individuals can move back and forth. We designate the areas as Area 1 and Area 2. Thus, the system can be depicted as a reaction-diffusion process in which reactions (6) and (7) occur independently inside both areas, while individuals can traverse between Area 1 and Area 2. An illustration is shown in Fig. 2

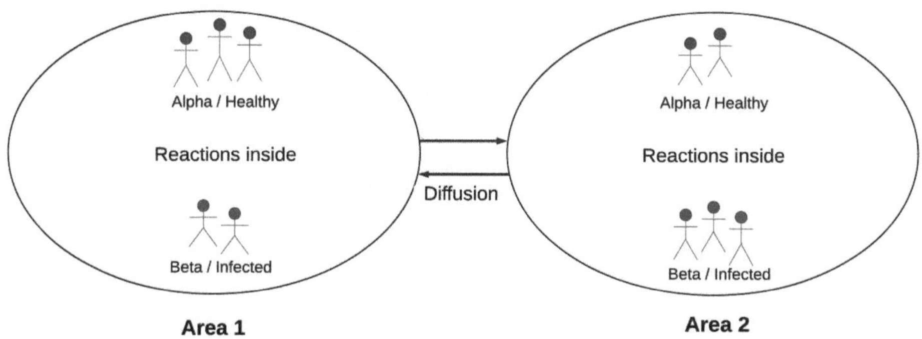

Fig. 2. Illustration of RDME in metapopulation model.

The process is a random process and can be modeled by the RDME consisting of two reactions with two compartments [14]. Let α_1 be the number of healthy individuals in area 1, β_1 the number of infected individuals in area 1, α_2 the number of infected healthy in area 2, β_2 the number of infected individuals in area 2. The reactions R_1 through R_4 in table 1 reflect (6) and (7) inside both areas, while R_5 through R_8 represent diffusion between areas.

For experimental purpose, we assume that initially, there are 100 healthy individuals and 90 infected individuals in area 1. Similarly, in area 2 there are 90 healthy and 100 infected individuals. The initial state vector is $[100, 90, 90, 100]^T$, and we set the reaction parameters $c_1 = .80$ and $c_2 = .30$, and the diffusion parameters $D = 1.0$ and $l = 10$, resulting $d = D/l^2 = 1.0/100 = .01$.

We use FORTRAN and OpenMP to implement the SSA algorithm for the metapopulation model employing the RDME. We use a do-loop of 1, 000, 000, 000 iterations. The Alabama SuperComputer (ASC), a cluster of high performance supercomputers, is used to simulate the trajectories. Table 2 shows the computational time for computing 1, 000, 000, 000 individual simulations of the SSA algorithm. We are able to analyze the speed-up and the parallel efficiency of the SSA algorithm. Figure 3 shows the analysis of the parallel implementation.

Table 1. Reactions, propensities and state change vector of metapopulation model

	Reactions	Propensities	State change vector
R_1:	$\beta_1 \xrightarrow{c_1} \alpha_1$	$c_1 \times \beta_1$	$[1, -1, 0, 0]$
R_2:	$\beta_1 + \alpha_1 \xrightarrow{c_2} 2\beta_1$	$c_2 \times \beta_1 \times \alpha_1$	$[-1, 1, 0, 0]$
R_3:	$\beta_2 \xrightarrow{c_1} \alpha_2$	$c_2 \times \beta_2$	$[0, 0, 1, -1]$
R_4:	$\beta_2 + \alpha_2 \xrightarrow{c_2} 2\beta_2$	$c_2 \times \beta_2 \times \alpha_2$	$[0, 0, -1, 1]$
R_5:	$\alpha_1 \xrightarrow{d} \alpha_2$	$d \times \alpha_1$	$[-1, 0, 1, 0]$
R_6:	$\beta_1 \xrightarrow{d} \beta_2$	$d \times \beta_1$	$[0, -1, 0, 1]$
R_7:	$\alpha_2 \xrightarrow{d} \alpha_1$	$d \times \alpha_2$	$[1, 0, -1, 0]$
R_8:	$\beta_2 \xrightarrow{d} \beta_1$	$d \times \beta_2$	$[0, 1, 0, -1]$

Table 2. Computational time for parallel computing of metapopulation model

	1,000,000,000 runs
1 Core:	42.46 sec
2 Cores:	20.27 sec
4 Cores:	10.29 sec
8 Cores:	5.93 sec
16 Cores:	2.04 sec

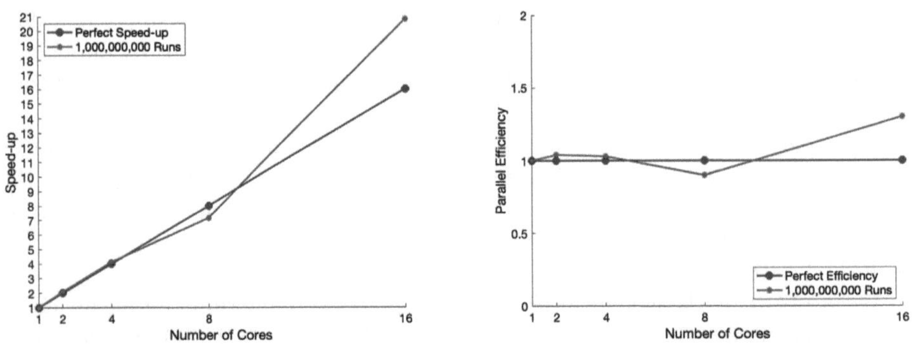

(a) Computational speed-up (b) Parallel efficiency

Fig. 3. Analysis of parallel implementation in metapopulation model

4.2 Birth-Death Process

The birth-death process is a fundamental stochastic model in molecular biology, used to model the dynamic behavior of biological systems where entities are created (birth) and removed (death) over time. It is particularly useful in modeling population dynamics, biochemical reactions, queuing systems, and epidemiological processes. We work with a well-studied birth-death process, which can be seen as a simplified model of the production and degradation of a single molecular species [17]. Consider the model for the production and degradation of a specific protein *A*. Protein molecules can be independently degraded at a constant rate c_1 and be produced at a constant rate c_2. The reaction scheme is given by

$$A \xrightarrow{c_1} \emptyset, \tag{8}$$

$$\emptyset \xrightarrow{C_2} A. \tag{9}$$

We consider two neighboring cells (i.e. compartments) where protein molecules can move back and forth. We label the cells as Cell 1 and Cell 2. Thus, the system can be represented as a reactiondiffusion process in which reactions (8) and (9) occur independently inside both cells, while molecules can move between Cell 1 and Cell 2. We model the scheme with RDME consisting of two reactions with two compartments illustrated in Fig. 4. Let A_1 be the number of protein molecules in cell 1, A_2 the number of protein molecules in cell 2. The reactions R_1 through R_4 in Table 3 reflect (8) and (9) inside both cells, while R_5 through R_6 represent movement between cells.

Table 3. Reactions, propensities and state change vector of birth-death model

	Reactions	Propensities	State change vector
R_1:	$A_1 \xrightarrow{c_1} \emptyset$	$c_1 \times A_1$	$[-1, 0]$
R_2:	$\emptyset \xrightarrow{c_2} A_1$	c_2	$[1, 0]$
R_3:	$A_2 \xrightarrow{c_1} \emptyset$	$c_1 \times A_2$	$[0, -1]$
R_4:	$\emptyset \xrightarrow{c_2} A_2$	c_2	$[0, 1]$
R_5:	$A_1 \xrightarrow{d} A_2$	$d \times A_1$	$[-1, 1]$
R_6:	$A_2 \xrightarrow{d} A_1$	$d \times A_2$	$[1, -1]$

Consider that initially, there are 100 protein molecules in cell 1, and 100 protein molecules in cell 2. The initial state vector is $[100, 100]^T$, and we set the reaction parameters $c_1 = .60$ and $c_2 = .20$, and the diffusion parameters $D = 1.0$ and $l = 10$, resulting $d = D/l^2 = 1.0/100 = .01$.

We employ parallel processing for the birth-death model using RDME. We use a do-loop of 1, 000, 000, 000 iterations. Table 4 shows the computational time for computing 1, 000, 000, 000 individual simulations of the SSA algorithm. The speed-up and the parallel efficiency of the SSA algorithm are shown in Fig. 5.

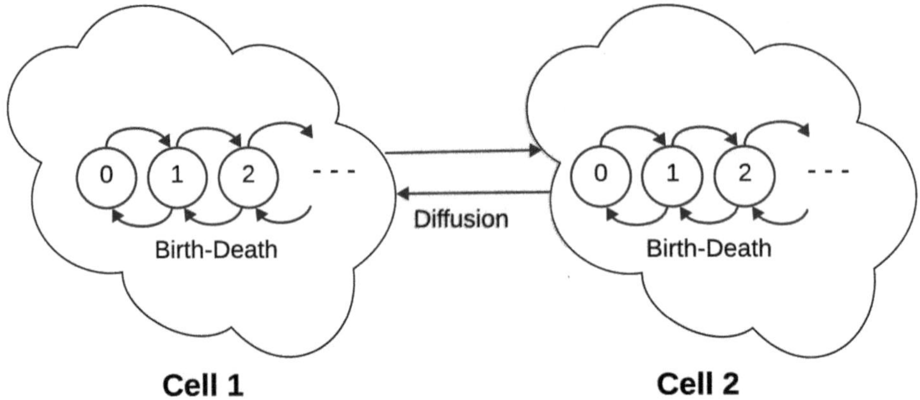

Fig. 4. Illustration of RDME in birth-death model

Table 4. Computational time for parallel computing of birth-death model

	1,000,000,000 runs
1 Core:	39.66 sec
2 Cores:	18.62 sec
4 Cores:	9.04 sec
8 Cores:	4.64 sec
16 Cores:	3.08 sec

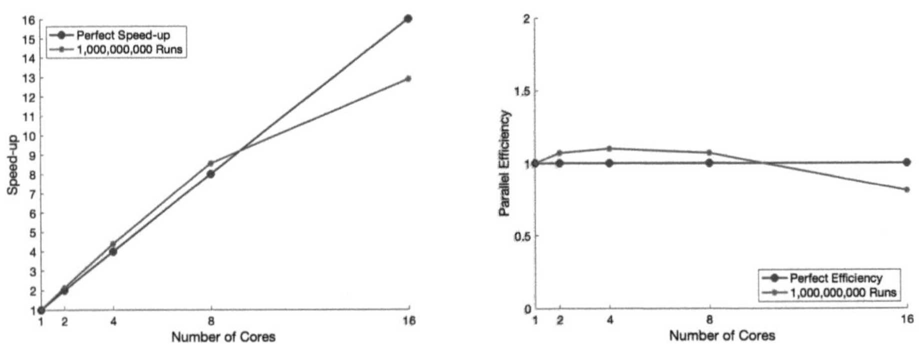

(a) Computational speed-up (b) Parallel efficiency

Fig. 5. Analysis of parallel implementation in birth-death model

5 Discussions and Conclusion

We have employed the RDME formulation in two systems biology models that encompass both reaction and diffusion processes. We used the SSA algorithm to simulate stochastic trajectories. Implementing the SSA in the RDME is computationally expensive. We use parallel processing to reduce the computational time of simulating the biological systems. Tables 2 and 4 demonstrate that computing time reduces with the utilization of additional cores. In Fig. 3, the speed-up for the metapopulation model is superlinear with 2, 4, and 16 cores, but sublinear with 8 cores. This signifies that the efficiency exceeds 1 for 2, 4, and 16 cores, while it drops below 1 for 8 cores. Greater cache efficiency, less memory contention, and parallelized random number generator contribute to the superlinear speed-up. However, sublinear speed-up occurs due to thread synchronization overhead, memory bandwidth saturation, and computational load imbalance. For the birth-death model, the speed-up is shown in Fig. 5. Superlinear performance is attained with 2, 4, and 8 cores, demonstrating efficiencies greater than 1, but shifts to sublinear speed-up with 16 cores, indicating a decline in efficiency below 1. These findings underscore the significance of parallel computing in addressing the computational requirements of systems biology models that incorporate stochastic reaction-diffusion processes. Future study will focus on using parallel computing approaches to more intricate biological systems. The emphasis will be on enhancing efficiency through advanced parallelization strategies or hybrid computing approaches to better handle large-scale biological simulations.

References

1. Anderson, R.M., May, R.M., Anderson, B.: Infectious Diseases of Humans: Dynamics and Control. Oxford University Press, Oxford (1992)
2. Bernstein, D.: Simulating mesoscopic reaction-diffusion systems using the Gillespie algorithm. Phys. Rev. E **71**(4), 041103 (2005)
3. Burrage, K., Burrage, P., Jeffrey, S., Pickett, T., Sidje, R., Tian, T.: A grid implementation of chemical kinetic simulation methods in genetic regulation. In: Proceedings of the APAC03 Conference on Advanced Computing, Grid Applications and eResearch (2003)
4. Colizza, V., Pastor-Satorras, R., Vespignani, A.: Reaction–diffusion processes and metapopulation models in heterogeneous networks. Nat. Phys. **3**(4), 276–282 (2007)
5. Dobrzyński, M., Rodríguez, J.V., Kaandorp, J.A., Blom, J.G.: Computational methods for diffusion-influenced biochemical reactions. Bioinformatics **23**(15), 1969–1977 (2007)
6. Doubrovinski, K., Howard, M.: Stochastic model for Soj relocation dynamics in bacillus subtilis. Proc. Natl. Acad. Sci. **102**(28), 9808–9813 (2005)
7. Gardiner, C., McNeil, K., Walls, D., Matheson, I.: Correlations in stochastic theories of chemical reactions. J. Stat. Phys. **14**, 307–331 (1976)
8. Gillespie, D.T.: A general method for numerically simulating the stochastic time evolution of coupled chemical reactions. J. Comput. Phys. **22**(4), 403–434 (1976)
9. Gillespie, D.T.: Exact stochastic simulation of coupled chemical reactions. J. Phys. Chem. **81**(25), 2340–2361 (1977)
10. Gillespie, D.T.: A rigorous derivation of the chemical master equation. Physica A **188**(1–3), 404–425 (1992)
11. Higham, D.J.: Modeling and simulating chemical reactions. SIAM Rev. **50**(2), 347–368 (2008)

12. Hill, M.D., Marty, M.R.: Amdahl's law in the multicore era. Computer **41**(7), 33–38 (2008)
13. Metzler, R.: The future is noisy: the role of spatial fluctuations in genetic switching. Phys. Rev. Lett. **87**(6), 068103 (2001)
14. Rahman, M.M., Sidje, R.B.: A study of a metapopulation model using the stochastic reaction diffusion master equation. In: Han, H., Baker, E. (eds.) Next Generation Data Science. pp. 242–253. Springer Nature Switzerland, Cham (2024)
15. Smoluchowski, M.: Mathematical theory of the kinetics of the coagulation of colloidal solutions. Z. Phys. Chem. **92**, 129–168 (1917)
16. Stundzia, A.B., Lumsden, C.J.: Stochastic simulation of coupled reaction–diffusion processes. J. Comput. Phys. **127**(1), 196–207 (1996)
17. Wilkinson, D.J.: Stochastic modeling for quantitative description of heterogeneous biological systems. Nat. Rev. Genet. **10**(2), 122–133 (2009)

Prediction and Analysis of Postpartum Depression with Chronic Diseases as Risk Factors

Febina M. Rajeesh and Waqar Haque(✉)

University of Northern British Columbia, Prince George, BC V2N 4Z9, Canada
{muhamed,waqar.haque}@unbc.ca

Abstract. Postpartum Depression (PPD) is a complex and prevalent mental health condition, often challenging to diagnose and treat effectively. While PPD has been widely studied in the context of traditional risk factors, the research linking chronic diseases to mental health issues is limited. Predicting PPD by utilizing the power of machine learning algorithms can lead to timely intervention and management of the condition. In this paper, we present results obtained from various predictive machine learning models to forecast the likelihood of depression for mothers who suffer from one or more chronic diseases. All models demonstrate a close alignment in their SHAP analysis in identifying key predictors. The results also highlight how chronic diseases could potentially interact with other common risk factors and increase the likelihood of PPD. The models consistently identified chronic diseases, asthma, and anemia as key predictors along with other interacting features. The results demonstrate the potential of machine learning as a screening tool to improve diagnostic precision and support personalized care for enhanced quality of life. An interactive diagnostic screening tool based on these models was also developed.

Keywords: Postpartum Depression · Machine Learning · Predictive Modelling

1 Introduction

Processing large volumes of complex health data with traditional methods often takes too long to reveal valuable insights. Recently, Machine Learning (ML) techniques have gained increased attention in extracting hidden patterns, relationships within datasets, diagnosing health conditions, predicting outcomes, and other valuable information from the massive quantity of complex health data. ML-based systems can aid clinicians in making more accurate assessments and develop individualized treatment plans to improve the quality of healthcare services and patients' quality of life. Specifically, in mental health care and management advancements, the invaluable contributions of ML include detecting and diagnosing mental health conditions, improving treatment options and patient support, contributing to public health initiatives, and supporting research and administrative work in clinical settings. The global market for mental health apps is expected to grow from \$6.22 billion in 2023 to \$13.3 billion by 2028, with a compound

T. Aldwairi et al. (Eds.): BICOB 2025, CCIS 2535, pp. 127–141, 2025.
https://doi.org/10.1007/978-3-031-94039-2_11

annual growth rate (CAGR) of 16.6% [1]. This increase reflects the growing role of ML and digital technologies in mental health, enabling more personalized care, enhanced support, and improved accessibility through relevant applications.

1.1 Background

Maternal mental health poses a considerable global challenge. Nearly one in five women will encounter a mental health condition during pregnancy or within the year following childbirth; of those with perinatal mental health conditions, 20% will also be grappling with suicidal thoughts or engaging in acts of self-harm [2]. Unfortunately, postpartum psychiatric disorders are often disregarded and left untreated. Mothers struggling with mental health challenges may also experience exacerbated physical health issues, which does not only affect their mental health and emotional well-being but could also potentially impact the well-being of their infants. Timely intervention is important to prevent complications. Postpartum depression (PPD) does not go away on its own, and it is one of the leading causes of the annually reported infanticide incidence [3]. Recognizing key risk factors enables healthcare providers to identify women who may be at high risk of developing PPD, ensuring they receive medical interventions and monitoring to lower the persistence during motherhood [4].

Machine learning offers automated solutions, refining processes, enhancing predictive accuracy, and revealing complex patterns in data. The success of these techniques suggests enormous potential for aiding mental health decision-making and enhancing research and clinical tools [5]. Early intervention and targeted support could be facilitated for at-risk mothers by predictive modelling, potentially resulting in improved outcomes for both mothers and their infants.

1.2 Motivation

In 2019, almost one-quarter of mothers who recently gave birth in Canada reported conditions consistent with either PPD or an anxiety disorder [6]. Although the risk factors for PPD have been studied, many potentially critical clinical variables remain unexplored [7]. PPD is often linked to immediate hormonal changes after childbirth, as well as a history of depression and socio-demographic challenges. Previous research on mental illness has demonstrated that depression frequently co-occurs with chronic diseases [8, 9]. Further evidence suggests that women with chronic conditions such as diabetes, hypertension, heart disease, migraines, and other neurological disorders are at an increased risk for peripartum mental illness [10]. However, studies specifically investigating the relationship between chronic diseases and PPD are limited.

The work presented in this paper examines relationship between independent variables, including chronic disease features, and the target variable PPD using neural networks (NN), logistic regression (LR), support vector machines (SVM), and Random Forest (RF) models. This approach has proven to be very effective for analyzing PPD by focusing on prediction, optimizing risk assessment, and improving clinical decision-making [11, 12].

2 Literature Review

Previous studies in this field have primarily aimed at identifying risk factors and developing models to facilitate timely intervention and effective management of PPD. In this section, we provide an overview of the current knowledge, gaps, and advancements in predicting PPD and broader mental health.

2.1 PPD Predictive Analysis

A recent systematic review evaluated the risk of bias and applicability of PPD prediction models commonly used in screening and diagnosis [13]. A scoping review [14] on ML methods for predicting PPD highlights the need for further clinical research collaborations to fine tune these methods for improved prediction and treatment outcomes.

A two-year longitudinal study of psychometric testing and predictive validity of the Postpartum Depression Predictors Inventory-Revised (PDPI-R) tool was conducted [15]. The study identified various predictors of PPD, which included unwanted pregnancy, previous history of depression, life stress, self-esteem, social support, and marital satisfaction. The tool examined both prenatal and postnatal periods, with prediction accuracy ranging from 64% to 81%. While the findings support routine screening to assess women's susceptibility to PPD and provide targeted support and intervention, generalizability and accuracy interpretation is limited.

Several ML approaches were presented for predicting risk level of PPD [16] using data from 686 mothers in Sri Lanka. The data was collected within six months of childbirth and the questionnaire included multiple factors such as physical disabilities, history of drug abuse, life stressors, social support, socioeconomic status, family problems, marital status, and education level. The models were built using an Adaptive Neuro Fuzzy Inference System with Genetic Algorithm (ANFIS-GA), ANFIS, and Feed For- ward Artificial Neural Network (FFANN), SVM, and RF. Among these models, the FFANN model with 25 epochs exhibited superior performance compared to other models. Despite promising findings, the study did not elaborate on the feature selection process for the predictive models or identify risk factors for PPD which could have provided valuable insights into the underlying causes of PPD.

Several data mining algorithms and ensemble techniques were used to assess the influence of quality of life on the probability of PPD [17]. The study examined 96 women who had given birth within the past year, irrespective of their prior depression history, and considered both urban and rural areas. The selected models achieved an accuracy of 84–94%. Findings revealed associations between unplanned pregnancy, spousal support, delivery method, and urban residence. The study revealed that women living in rural areas have had a lower prevalence of PPD, indicating a positive impact of lifestyle and living conditions. The limitation includes a small participant pool and limited sampling area.

A study was conducted to find the most effective ML algorithms for early prediction of PPD [18]. The study examined 28,755 records from the PRAMS dataset spanning from 2012 to 2020 and utilized nine ML algorithms to predict PPD considering health factors such as pre-pregnancy BMI, nutrition status, and health problems while ignoring demographic and socioeconomic factors. The analysis of different models revealed that SVM performed the best overall achieving an accuracy of 74.4%. SVM also had the highest precision and F1 score, indicating its effectiveness in making accurate predictions. However, the state-specific nature and limited diversity of the dataset may affect the model's accuracy when applied to different populations.

RF was identified as most effective in predicting PPD, followed by the SVM in a study that implemented nine ML algorithms utilizing the PRAMS 2012–2013 dataset [12]. Random Down-Sampling and Synthetic Minority Over-Sampling Techniques (SMOTE) were used to address the imbalance. Model performances were evaluated using 10-fold cross-validation, with AUC as the performance metric for comparison. Younger age, lower education, small-for-gestational-age infants, smoking reduction during pregnancy, and unmarried status were associated with a higher likelihood of PPD. Other important key predictors were stress during pregnancy, prior depression, weeks of breastfeeding, income, dental hygiene, and infant gender. Self-reporting bias and the likelihood of limiting generalizability due to missing data were recognized as limitations.

A study was done to diagnose PPD in women and develop appropriate treatment plans using demographic, depression, and pregnancy survey data [19]. The ML algorithms were implemented to examine several stressors and found that women with depression before getting pregnant were four times more likely to have PPD. Furthermore, the highest prevalence of PPD was observed in women who were younger, unmarried, had a lower income, had low education, or had a baby born less than 37 weeks. As this study focused on survey data, the clinical variables were not used to predict PPD.

2.2 PPD Risk Factors: Chronic Illness

Chronic illnesses make patients more likely to have or develop mental health problems. For instance, a longitudinal investigation within a large prospective cohort identified relationship between Diabetes Mellitus in Pregnancy (DMP) and perinatal depression [19]. The findings revealed a significant positive association between various types of Diabetes Mellitus and postpartum depression. The study acknowledges limitations which may impact generalizability, and possible biases due to sample characteristics such as higher education levels.

The prevalence of postpartum depression was high in an institutional-based cross-sectional study conducted in Ethiopia [20]. This study uses the SPSS statistical package to conduct a bivariate analysis for finding the first association between the independent and dependent variables. Among the 408 participants, 30 (7.4%) had chronic medical illnesses, 33.8% had PPD, and those who had other chronic physical diseases were 7.7 times more likely to develop PPD. In addition, unstable marital status, infant death incident, unplanned pregnancy, and age demonstrated increased probability for PPD. Even though this research explored chronic physical disease, the specific chronic illnesses belonging to the risk factors of PPD were not identified.

A variety of other limited studies to determine the role of single or multiple chronic diseases in PPD have also been conducted. It is established that women with chronic illnesses had a chance of having PPD, and those having more than one chronic physical illness had an increased risk for PPD.

2.3 Summary

While research has shed some light on the relationship between chronic illnesses and PPD, it remains a relatively understudied area. Significantly, the extant body of literature has not conclusively identified specific chronic diseases that can be classified as predictors for the onset of PPD. A high prevalence of depression among individuals with chronic illnesses such as coronary heart disease, diabetes, and hypothyroidism was indicated in research focusing on mental health disorders and chronic illness. Despite a clear connection between depression and certain chronic illnesses, the exact underlying mechanisms remain unknown.

3 Our Methodology

Predictive modeling anticipates future outcomes by utilizing historical and current data and applying advanced statistical techniques and machine learning algorithms. It identifies patterns and relationships within the data during training, establishes the relationship between independent and predicted variables, and provides meaningful findings. Predictive models play an important role in clinical decision-making, helping to improve personalized and targeted treatments [21]. Our research methodology consists of several components ranging from data collection to visual analytics (Fig. 1).

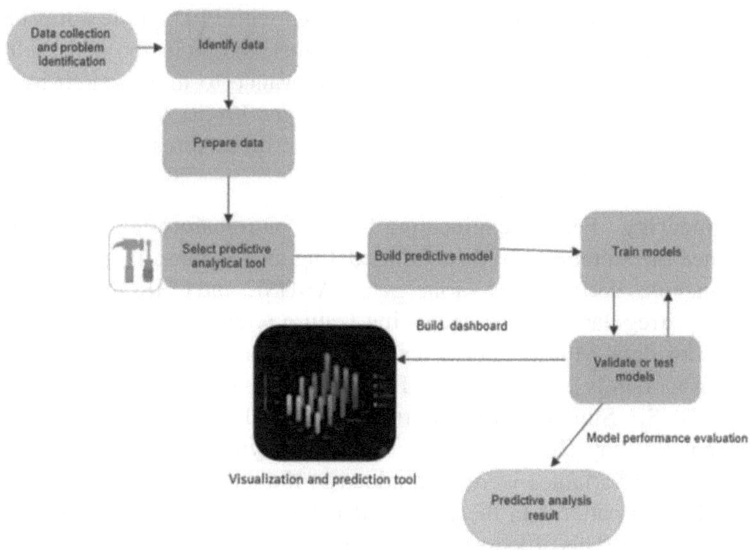

Fig. 1. Research Methodology

3.1 Data Collection

The data used for this research was obtained from CDC-PRAMS (Centers for Disease Control Pregnancy Risk Assessment Monitoring System). The data was collected using a mixed-mode surveillance system and represents 202,745 mothers surveyed across all US states from 2016 to 2020. It includes maternal demographic factors, chronic disease, postpartum, gestational, and obstetric variables.

3.2 Data Preparation

The raw data was cleansed, transformed, and organized to make it suitable for analysis. Initial data preprocessing also included broader tasks like feature engineering, data integration, and handling imbalances in labeled data. Imputation was applied to handle missing data due to the skipping pattern, while non-response missing data was addressed through row deletion because the assumption that similar mothers would respond the same way may not always be valid, and the imputation process is statistically complex.

Feature Selection. Feature selection is a vital step in machine learning, aiming to choose a subset of the most relevant features (input variables) from the original set of features. Its purpose is to boost the performance of the model, mitigate overfitting, and enhance the interpretability of the model. Considering our research question to determine whether chronic diseases are among the risk factors and identifying those which are most likely to trigger PPD, the input variables included all chronic diseases available in the dataset in addition to the conventional risk factors associated with PPD and features identified from feature selection node in SPSS modeler. Further, the inclusion and exclusion of variables was based on data completeness and relevance to the research hypothesis. Participant records were filtered to include mothers aged 17–50. Variables critical to the study were prioritized, while redundant or less impactful features were removed using the feature selection algorithm in SPSS modeler.

Feature Engineering. Feature engineering focuses on transforming and creating features to enhance the suitability of data for modeling and increase its predictive power. This process includes deriving new variables like ABUSE, and CHRONIC from existing ones, correcting irregularities, and enhancing feature relevance and meaningful transformations of raw data. Multiple abuse features, including PAD6HUS (from husband), PAD_FAM (from family), and PAD_OTH (from others), are aggregated to create a new feature, ABUSE. Similarly, a transformed feature CHRONIC, was created to indicate the presence of any chronic disease, including thyroid dysfunction, diabetes, or asthma. The derived variables were then added to the dataset and these new features helped reveal hidden patterns in the data, enhancing the accuracy and performance of the model's prediction. This process resulted in a final dataset comprising 23,855 records and 42 features. A nurse practitioner reviewed and confirmed the selection of these final variables, ensuring the clinical relevance and validity of the data used for analysis.

Data Encoding and Scaling. Data transformations and scaling are essential steps in data preparation for ML models. This is important to address issues arising from disparate data ranges and categorical representations, which can otherwise result in biased predictions or inefficiencies in algorithm performance. Encoding involves converting categorical variables into numerical formats that algorithms can interpret. Tree-based algorithms, such as RF, do not require encoding for categorical data because they natively handle such data due to their tree-splitting mechanism, and they can evaluate categories directly without assuming any data type-specific relationships. In this study, pre-processed data was scaled for features such as maternal and paternal race, and month of birth for LR, NN, and SVM models. Similarly, scaling ensures that numerical variables are standardized to a consistent range or distribution, improving model convergence and performance. Variables such as infant age were scaled for NN and SVM algorithms because of sensitivity to feature magnitude or distribution.

3.3 Challenges

Conducting thorough exploratory data analysis (EDA) is essential to identify imbalances in datasets, which can severely affect the performance of ML models. Unbalanced data can skew the model predictions toward the majority class. SMOTE was applied in this research to handle the imbalances in data. SMOTE was used to balance the dataset and improve model's ability to generalize across classes, enhancing overall predictive performance. For each minority class sample, SMOTE finds the nearest neighbors using Euclidean distance in feature space, generates synthetic points between them, and repeats this process to balance the dataset without duplicating data.

4 Experiments and Results

The source data was integrated into a database using SQL Server Management Studio (SSMS) 16.0. Subsequently, SPSS Modeler 18.4 was used to refine the selection of relevant variables in consultation with medical personnel. The preprocessed data was used to develop the model and generate results in Python. The model underwent training, testing, and validation stages to ensure its accuracy and reliability in predicting outcomes related to the research objectives. The dataset was partitioned into training and testing sets with a 70:30 ratio, respectively. Machine learning models were trained to learn the underlying patterns and associations between the input features and the target variable. The models were retrained by tuning the parameters and optimizing them to enhance the performance of each model. Also, cross-validation was applied to reduce the risk of overfitting. A stratified 5-fold cross-validation was used to validate how well models generalize by ensuring that folds maintain the same proportion of each class as the original dataset, which is important given the imbalanced nature of data. Dataset splits into 5 balanced folds, iteratively trains on 4 folds while testing the remaining, ensures stability by evaluating different subsets, and averages performance metrics across all 5 folds for a reliable estimate of model performance across multiple splits.

4.1 Evaluation Metrics and Selected Models

The trained model's effectiveness was assessed by measuring accuracy and F1 score, the harmonic mean of *precision* and *recall*. The accuracy was determined by dividing the number of correct predictions by the total number of predictions. *Precision* is the ratio of correctly predicted positive observations to the total predicted positives (true and false). *Recall* (or sensitivity) is the ratio of correctly predicted positive observations to the sum of true positives and false negatives. The F1 score then helps to balance the trade-off between false positives and false negatives, ensuring a thorough assessment of the model's predictive capabilities. Together, accuracy and F1 score provide a complete view of the model's effectiveness in accurately classifying observations.

Four models were selected for this study. First, a Multi-Layer Perceptron (MLP) Neural Network (NN) Classifier was chosen because it balances complexity and flexibility, making it a good fit for different binary classification tasks. Hyperparameters, the configurable settings were optimized using RandomizedSearchCV which picks and tests a set number of random combinations. The best parameters were identified through random sampling and evaluating a subset of possible combinations. The model gave 76% accuracy and an F1 score of 77%. Second, Logistic Regression (LR) was chosen for its ability to predict the binary target using a logistic function. A GridSearchCV approach was used to optimize the hyperparameters of the LR model, resulting in the selection of a maximum number of iterations as 5000 and the solver method as 'liblinear'. Using 5-fold stratified cross-validation, the search was performed in parallel to enhance computational efficiency. The accuracy and F1 scores of 89% and 88%, respectively, were achieved. Third, the Random Forest (RF) model was used with 300 trees in the forest with a leaf node, two minimum samples to split a node with no limit to the depth of tree, and without setting bootstrap sampling. This model yielded an accuracy and F1 score of 83% and 82%, respectively. Fourth, Support Vector Machine (SVM) was optimized through RandomizedSearchCV and using the pipeline which selected the Radial Basis Function (RBF) as the kernel and achieved accuracy and F1 scores of 83% and 80%. For all models, the closeness of accuracy and F1 scores suggest that the models are not biased toward one class over the other, and there is a good balance between precision and recall.

4.2 Post-training Analysis

This section identifies key predictors, focusing on the relationship between chronic diseases and the risk of PPD. SHAP (Shapley Additive exPlanations), a game theoretic approach, was utilized in the post-training model analysis. Since SHAP can be model-specific or agnostic, it has the flexibility to adapt to various ranges of models. The SHAP value was calculated to interpret the impact of the input features on the target PP_DEPRESS. Plots are based on the mean absolute SHAP values of each important feature. The beeswarm plots show a deeper insight into each value of every significant predictor, whether it has a positive or negative impact. Each dot corresponds to an individual prediction, shifting the color gradient from blue for lower feature values to red for higher ones. The positive SHAP value for a feature value indicates that it has a strong positive contribution to predicting PPD, and the negative value indicates the reverse. Generally, the feature significance is consistent across all models.

Neural Network. Figure 2 shows the key features identified by the NN model using SHAP. The bars show the absolute SHAP value with more important features shown at the top. The feature CHRONIC (red-presence) had the highest SHAP value of 1.33, followed by several other chronic diseases: anxiety, anemia, and asthma. Other chronic diseases (heart disease, epilepsy, thyroid dysfunction, and pre-pregnancy diabetes) are displayed with lower SHAP value for their presence.

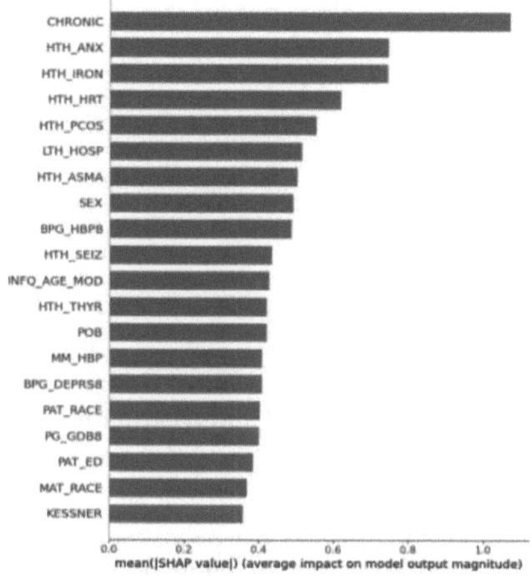

Fig. 2. Feature significance (NN)

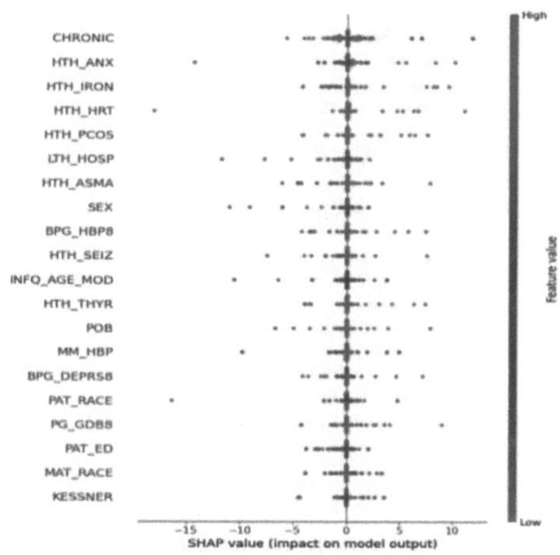

Fig. 3. Feature Impact and Distribution (NN)

Figure 3 provides the corresponding SHAP analysis for each feature. Length of stay at the hospital, high blood pressure, race, education, and Kessner Index are also flagged as predictors though with lower importance. The Kessner Index measures the adequacy of prenatal care by evaluating when prenatal care began, total number of visits, and the length of pregnancy.

Logistic Regression. Figure 4 shows the important predictors of the PPD when using LR. The feature CHRONIC comes after paternal and maternal race. The CHRONIC and MH_PGDX8 (Health problem during pregnancy-depression) got the same SHAP value of 0.29. However, it is still identified as a significant contributor in predicting PPD.

The corresponding SHAP analysis across each feature is shown in Fig. 5. Taking the feature CHRONIC, the red (1-presence of chronic disease) falls under the positive SHAP value section, indicating that the mother with chronic disease had a higher chance of PPD risk. In addition to the presence of chronic disease, lower maternal and paternal education level, younger age of mother, fewer prenatal care visits, lower Kessner Index, higher stress level, marital status, epilepsy and history of depression are identified as strong contributors for predicting the PPD risk.

Random Forest. The key features identified by the RF model using SHAP are shown in Fig. 6. Like other models, the feature CHRONIC is indicated as one of the key predictors of PPD. In addition, other similar features such as depression during pregnancy, stress level, maternal age group, maternal and paternal education and race, asthma and anemia are found in the top ten predictors.

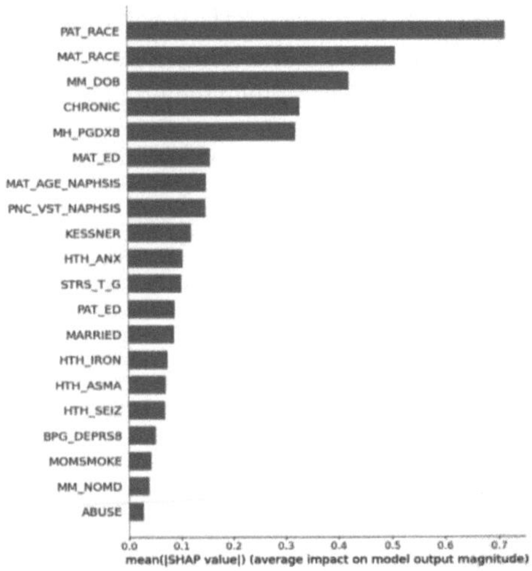

Fig. 4. Feature significance (LR)

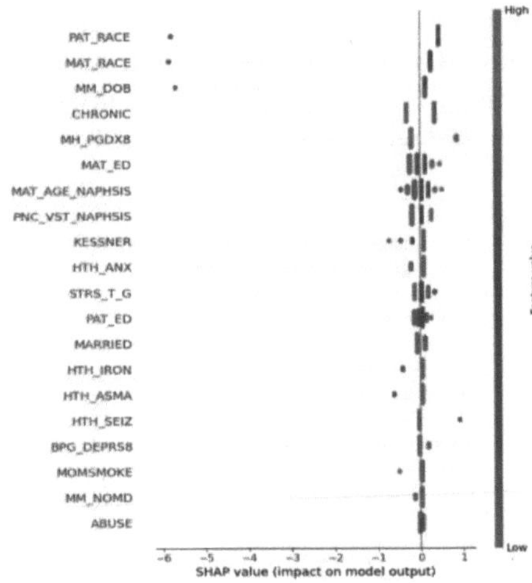

Fig. 5. Feature Impact and Distribution (LR)

The SHAP summary plot in Fig. 7 shows the detailed effect of each feature value in predicting PPD, once again highlighting that the feature CHRONIC has a positive contribution when present, consistent with both the LR and NN models. Notably, not breast feeding emerges as an additional vital contributor in importance feature.

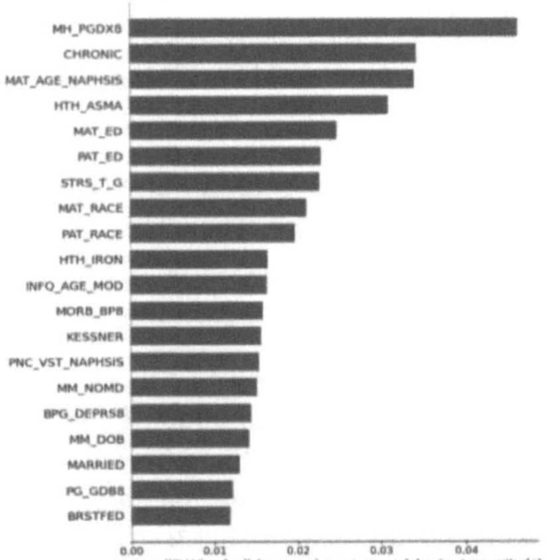

Fig. 6. Feature significance (RF)

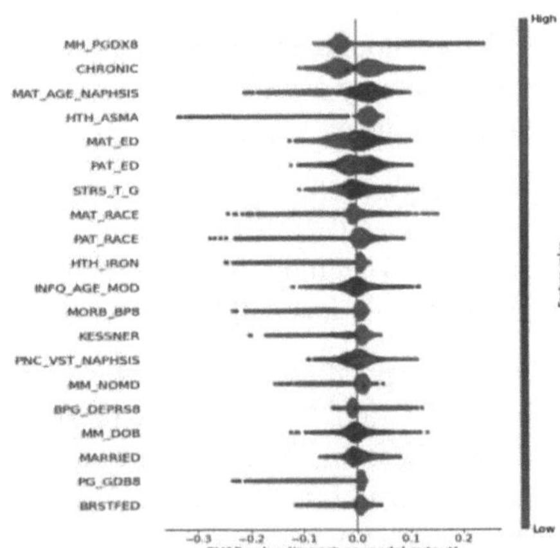

Fig. 7. Feature Impact and Distribution (RF)

Support Vector Machine. Figure 8 illustrates the key features identified by the SVM model using SHAP. The feature CHRONIC was at the top, with a SHAP value of more than 0.90, followed by other selected chronic diseases, such as anxiety, anemia, and asthma. Further down, thyroid dysfunction is also identified as an important feature in predicting PPD.

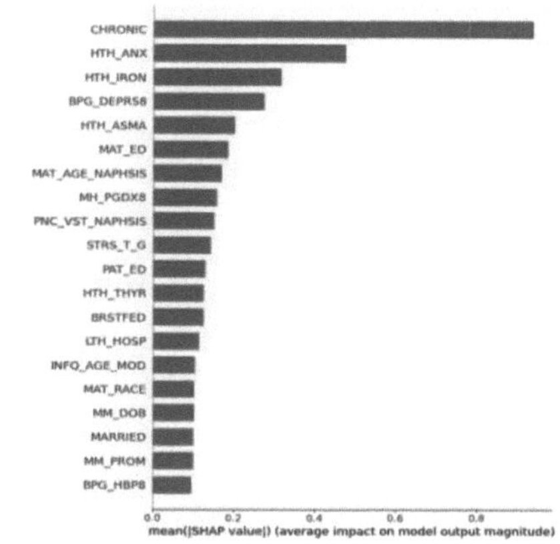

Fig. 8. Feature significance (SVM)

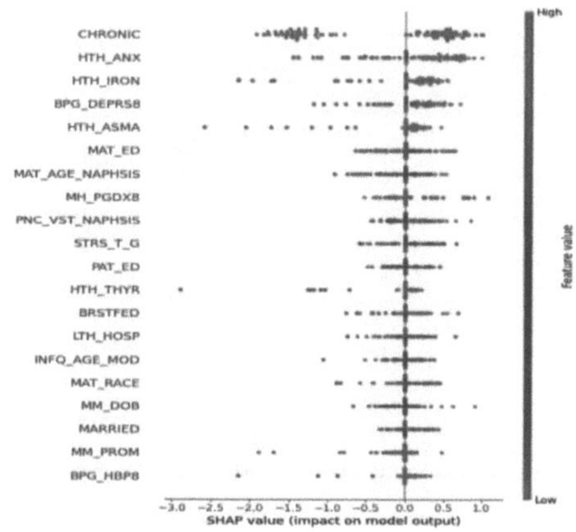

Fig. 9. Feature Impact/Distribution (SVM)

The beeswarm plot in Fig. 9 shows a detailed overview of how individual feature values affect the target. The distribution reveals that when the binary feature CHRONIC is 1, it gives a positive SHAP value, thus contributing positively to the PPD prediction. Consistent with the LR model, the mother's depression during pregnancy (MH_PGDX8) holds similar significance in predicting PPD. In contrast to LR, marital status (MARRIED) shows mixed values in both directions, making it difficult to determine which category of this feature is a vital contributor.

Summary: Post-training Analysis. CHRONIC, asthma, anemia, paternal education, maternal race, and history of depression are consistently labeled as key features across all models with comparable SHAP values. Additionally, marital status, depression during pregnancy, infant age, maternal and paternal education, Kessner Index, anxiety, stress level, maternal and paternal race, and maternal age were identified as key features among at least three out of the four models. Only a few additional vital features emerged uniquely in each model Notably, the features married displays different SHAP values indicating variability in its influence across models. The minor differences are likely due to each model's handling of nonlinearities and interactions. However, the SHAP values for the presence of individual chronic diseases were low, indicating a need for an interaction analysis. Such patterns often occur when models see these features as part of broader interactions rather than as strong individual predictors.

4.3 Identifying and Analyzing Interactions Among Key Predictors

The low SHAP values for individual chronic diseases indicated the likelihood for these features to contribute more significantly through interactions rather than as standalone predictors. New features were generated to assess the combined effect of features identified as important through SHAP analysis. Each new feature was given a meaningful

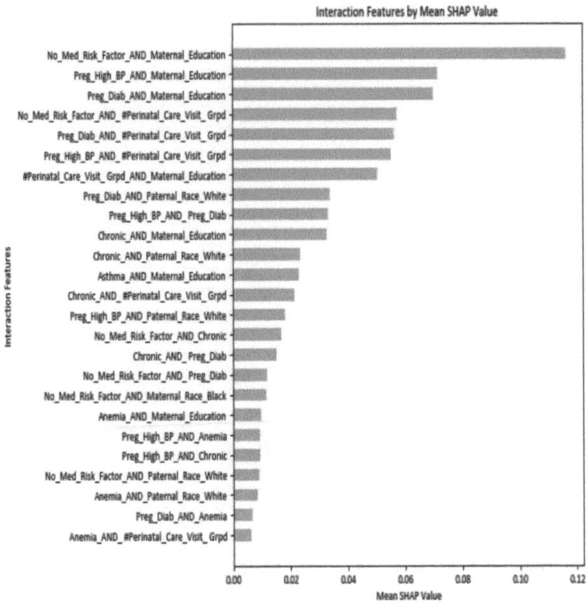

Fig. 10. Interaction Feature significance (NN)

name and added to the original dataset as a new column. Only those identified as significant by SPSS Modeler were retained. Finally, the enhanced dataset was trained and tested using the interaction features.

Figure 10 illustrates the key interaction features of the NN model identified using SHAP. While the absence of other medical risk factors is significant with the interaction of feature CHRONIC, the presence of chronic disease appears as a key predictor for PPD when combined with factors such as education level, high blood pressure, gestational diabetes, and race.

The significance of interaction features by other models exhibited similar observations. For instance, LR showed paternal education and maternal education at the top. The analysis found a moderate interaction between the CHRONIC feature and the maternal age group. RF identified paternal education without medical risk factors as the top predictor of PPD, which SVM also ranked among the top contributors. Both RF and NN models emphasize the combined significance of education and chronic disease as a key predictors of PPD. Additionally, chronic disease and maternal age group interaction is a predictor that aligns with LR results. Disease variable, asthma interacts with maternal age, paternal education, and infant age. For SVM, prenatal disease variables such as gestational diabetes and high blood pressure were on the top, interacting with paternal education, maternal age, prenatal care visit, Kessner Index, and length of stay in the hospital.

In summary, interaction analysis captures the combined influence of multiple features and valuable insights into influential factors of PPD. Features such as 'No medical risk factors,' high BP, gestational diabetes, maternal age group, number of prenatal care visits, chronic disease, and maternal and paternal education frequently interact with other predictors, emphasizing their roles in PPD prediction modeling. Notably, maternal education appeared across all models, while pregnancy with high BP, and diabetes appeared in two models: NN, and SVM.

In the SVM model, features such as the Kessner Index (excluding its interaction with paternal education), infant sex, and length of hospital stay showed interactions with other variables and were uniquely significant to this model. In the LR model, interactions involving abuse were uniquely significant. Similarly, in the NN model, interactions involving anemia stood out. Interactions involving asthma were particularly notable in RF and NN models.

5 Data Visualization and Interactive Calculator

A dashboard provides visual representation of relationships and patterns within the data to gain insights into PPD statistics, including prevalence, correlations, trends, and gaps in PPD care. The data is segmented across various dimensions, including maternal age groups, prevalence, disease variables, maternal education groups, prenatal variables, and infant factors over the years. An enhanced machine-learning-based tool to predict the likelihood of PPD as a percentage, provides a streamlined and interactive interface for mental health assessment. Details of this interactive platform are beyond the scope of this paper.

6 Conclusion

Analyzing diverse data sources, predicting risk factors, providing patient-specific treatment, and monitoring symptoms enhances the quality of mental health care. Predicting PPD by utilizing the power of machine learning algorithms can lead to timely intervention and management resulting in better quality of life. The four ML models used in our research exhibited a range of performance scores, from NN achieving 76% accuracy and 77% F1 score to LR reaching 89% accuracy and 88% F1 score indicating good predictive capabilities of the models. LR's effectiveness with linearly separable data and resistance to overfitting with regularization likely gave a higher performance score. Also, RF's robust ensemble learning and SVM's effective margin optimization with RBF Kernel likely contributed to their superior performance.

All models demonstrate a close alignment in their SHAP analysis in identifying key predictors. The SHAP summary plots reveal similar positive and negative values for multiple key features across the models. Features such as CHRONIC, asthma, anemia, history of depression, paternal education, and maternal race are also consistently flagged as important. CHRONIC consistently comes among the top three predictors, supporting the research hypothesis that chronic diseases are a significant predictor of PPD. In addition, variables such as maternal age, education, race, marital status, depression during pregnancy, higher stress levels, and longer stay at the hospital, contributed to the prediction of PPD. Also, key predictors such as depression (during and before pregnancy), higher stress, and maternal age are consistent with other findings [12, 15, 19]. A few interaction features appeared as significant predictors as well. The interaction analysis revealed factors such as high blood pressure, gestational diabetes, maternal age group, number of prenatal care visits, chronic disease, and maternal and paternal education interacting with other predictors frequently. Chronic disease, an important predictor in three models (NN, LR, and RF), also frequently interacted with other features highlighting its vital role in predicting PPD.

Acknowledgements. We acknowledge the Pregnancy Risk Assessment Monitoring System (PRAMS) working group and the Centers for Disease Control (CDC) and Prevention for providing access to the PRAMS dataset for this analysis.

References

1. The Business Research Company, "Mental Health Apps Global Market Report," https://www.thebusinessresearchcompany.com/report/mental- health-apps-global-market-report#:~:text=Mental%20Health%20Apps%20Market%20Size,(CAGR)%20of%2015.4%25. Accessed 9 Nov 2024
2. WorldHealthOrganization (2022). https://www.who.int/news/item/19-09-2022-launch-of-the-who-guide-for-integration-of-perinatal-mental-health. Accessed 20 Mar 2024
3. "Postpartum depression," Wikipedia, the free encyclopedia. https://en.wikipedia.org/wiki/Postpartum_depression. Accessed Aug 2022
4. Šebela, A., Hanka, J., Mohr, P.: Etiology, risk factors, and methods of postpartum depression prevention. Ceska Gynekol. **83**(6), 468–473 (2018)
5. Shatte, B.R., Hutchinson, D.M., Teague, S.J.: Machine learning in mental health: a scoping review of methods and applications. Psychol. Med. **49**(9), 1426–1448 (2019)
6. StatCan, 2018/2019. https://www150.statcan.gc.ca/n1/daily-quotidien/190624/dq190624b-eng.html. Accessed 14 Sep 2022
7. Ross, L.E., Dennis, C.L.: The prevalence of postpartum depression among women with substance use, an abuse history, or chronic illness: a systematic review. J. Women's Health **18**(4), 475–486 (2009)
8. Hamed, F., Supriyanto, E., Osman, S., Khider, T.A.E.: Risk prediction of major depressive disorder using. In: 2020 3rd International Seminar on Research of Information Technology and Intelligent Systems (ISRITI) (2020)
9. Chen, R., et al.: 3-year risk prediction of coronary heart disease in hypertension patients: a preliminary study. In: 2017 39th Annual International Conference of the IEEE Engineering in Medicine and Biology Society (EMBC) (2017)
10. Brown, H.K., Qazilbash, A., Rahim, N., Dennis, C.L., Vigod, S.N.: Chronic medical conditions and peripartum mental illness: a systematic review and meta-analysis. Am. J. Epidemiol. **187**(9), 2060–2068 (2018)
11. Hochman, E., et al.: Development and validation of a machine learning-based postpartum depression prediction model: a nationwide cohort study. Depress. Anxiety **38**(4), 400–411 (2021)
12. Shin, D., Lee, K.J., Adeluwa, T., Hur, J.: Machine learning-based predictive modeling of postpartum depression. J. Clin. Med. **9**(9), 2899 (2020)
13. Qi, W., et al.: Predictive models for predicting the risk of maternal postpartum depression: a systematic review and evaluation. J. Affect. Disord. **333**, 107–120 (2023)
14. Saqib, K., Khan, A.F., Butt, Z.A.: Machine learning methods for predicting postpartum depression: scoping review. JMIR Mental Health **8**(11), e29838 (2021)
15. Ayhan, F., Akalin, A., Balsak, H., Erden, A.: Psychometric testing and the predictive validity of the postpartum depression predictors inventory-revised (PDPI-R): a longitudinal study with Turkish women. Sexual Reprod. Healthc. **40**, 10096 (2024)

16. Prabhashwaree, T., Wagarachchi, N.M.: Towards machine learning approaches for predicting risk level of postpartum depression. In: 6th SLAAI International Conference on Artificial Intelligence, Colombo, Sri Lanka (2022)
17. Baruah, S., Mazumder, P.: A community based study for early detection of postpartum depression using improved data mining techniques. In: 2021 IEEE International Conference on Computation System and Information Technology for Sustainable Solutions (CSITSS), pp. 1–7 (2021)
18. Paul, A., Pragada, S.D., Murthy, D.N., Shruthi, M.L.J., Gurugopinath, S.: Performance comparison of machine learning techniques for early detection of postpartum depression using PRAMS dataset. In: 15th IEEE International Conference on Computational Intelligence and Communication Networks (2023)
19. Natarajan, S., Prabhakar, A., Ramanan, N., Bagilone, A., Siek, K., Connelly, K.: Boosting for postpartum depression prediction. In: 2017 IEEE/ACM International Conference on Connected Health: Applications, Systems and Engineering Technologies (CHASE), pp. 232–240 (2017)
20. Bjorvang, R.D., Liakea, I., Carpentsier, B., Kozinsky, Z., Skalkidou, A., Fransson, E.: Association of diabetes mellitus in pregnancy and perinatal depression. Psychosom. Med. **86**, 52–58 (2023)
21. Kerie, S., Menberu, M., Niguse, W.: Prevalence and associated factors of postpartum depression in Southwest, Ethiopia, 2017: a cross-sectional study. BMC. Res. Notes **11**(623), 2018 (2018)
22. Kumar, V.: Predictive analytics: a review of trends and techniques. Int. J. Comput. Appl. **182**(1), 31–37 (2018)

Graph Neural Networks Based Explainability of Drug-Target Interactions

Chandra Mohan Dasari[1(✉)], Dheeraj Kodati[2], Nivesh Mittapally[3], A. Srinivas Reddy[4], and P. Krishna Reddy[3]

[1] Indian Institute of Information Technology, Sri City, Sri City, Andhra Pradesh, India
chandramohan.d@iiits.in
[2] Mahindra University, Hyderabad, Telangana, India
[3] International Institute of Information Technology, Hyderabad, Hyderabad, Telangana, India
[4] GenZ Technologies Pvt. Ltd., Hyderabad, Telangana, India

Abstract. Over a decade, efforts have been made to extend data science-based approaches to improve the process of drug-target interactions (DTIs) by modeling drugs and targets as graphs. From a drug discovery point of view, a pharmacophore, a subgraph of a drug candidate, is an essential structural and chemical feature necessary for interacting with a specific biological target. Effective identification of potential pharmacophores from a given candidate drug is a research issue. In the literature, efforts are being made to investigate machine learning and Graph Neural Network (GNN) based frameworks to identify the potential drug candidates for a given target. However, the problem of extracting the knowledge of the potential pharmacophore of the given candidate drug has not been explored. There is an opportunity to extract pharmacophores from drug candidates by exploiting the knowledge of potential subgraphs extracted through a trained GNN model. In this paper, we propose a two-phase GNN-based framework for identifying drug candidates and extracting potential pharmacophores from these candidates. The framework consists of two parts. First, we employ a GNN-based model to compute the affinity score for the given drug compound. Second, by employing the Monte Carlo Tree Search (MCTS) algorithm, the trained GNN is leveraged to extract potential subgraphs representing pharmacophores. The experimental results on the Davis, Kiba and Allergy datasets demonstrate the feasibility of the proposed approach to extract pharmacophores of candidate drugs with high performance. The predicted binding affinities of molecular subgraphs extracted from drug candidates are very similar to those of the corresponding drug candidates. The proposed approach helps explore the potential pharmacophore of the given candidate drug, which enhances the explainability of DTI to obtain deeper insights into crucial molecular interactions.

Keywords: Pharmocophores · Drug Candidates · Explainability · Connected Components · Binding Affinity

T. Aldwairi et al. (Eds.): BICOB 2025, CCIS 2535, pp. 142–162, 2025.
https://doi.org/10.1007/978-3-031-94039-2_12

1 Introduction

Identifying Drug-Target Interactions (DTIs) is an essential step in the drug discovery process. The DTI step involves discovering and validating interactions between chemical compounds (drug candidates) and biological macromolecules (target proteins). Traditional methods, while foundational, often fall short of capturing the nuanced and dynamic nature of biological interactions [21]. Research efforts are going on to extend pattern mining, and data science approaches to enhance the DTI identification process.

This graph-based representation of drugs and targets encapsulates the structural and functional properties of molecules, providing a more holistic view of potential interactions. Efforts are being made to extend machine learning and graph neural networks (GNNs) to identify potential drug candidates. From a drug discovery point of view, a pharmacophore, a subgraph of a drug candidate, is an essential structural and chemical feature necessary for interacting with a specific biological target. These features typically include hydrogen bond donors and acceptors, hydrophobic regions, and charged groups that are spatially arranged to facilitate binding to the target. Identifying potential pharmacophores from the candidate drug is important, as they form the basis for designing and optimizing drug candidates with high specificity and affinity for their targets.

The problem of extracting the knowledge of potential pharmacophores from candidate drugs remains largely unexplored. While existing GNN models can identify drug candidates that are likely to interact with a given target, they do not explicitly extract and elucidate the pharmacophoric subgraphs that underlie these interactions [7]. There is an opportunity to extract pharmacophores from drug candidates by exploiting the knowledge of potential subgraphs extracted through a trained GNN model. Notably, explainable AI techniques are increasingly being integrated into machine learning models to provide transparency and understanding of the decision-making processes [38]. The explainable AI techniques provide an opportunity to identify the portion of the candidate drug that causes DTI. In the context of DTI, extracting potential pharmacophores from a given drug candidate improves explainability in terms of identifying crucial molecular interactions that are causing the binding affinities between drugs and targets. By elucidating these interactions, researchers can gain a deeper understanding of the molecular mechanisms at play, enabling more informed drug design and development decisions. Identifying pharmacophores will improve the explainability of drug-target interactions, which helps bridge the gap between computational predictions and practical applications, ensuring that the identified interactions are statistically significant and biologically relevant.

In this paper, we propose a two-phase GNN-based framework [25,36] for identifying drug candidates and extracting potential pharmacophores from these candidates. The framework consists of two parts. First, we propose a two-channel DTI framework for identifying effective drug candidates. The model determines the binding affinity between the given target protein and drug candidate so that only potential drugs can be considered for extracting pharmacophores. Second, we propose a framework that employs the Monte Carlo Tree Searching (MCTS)

[4] methodology to enumerate subgraphs and identify potential subgraphs by leveraging the proposed GNN-based model to compute reward scores.

The key contributions are as follows:

- Two-channel DTI model for identifying potential drug candidates for a target protein.
- MCTS-based framework to extract potential subgraphs based on the trained DTI model to compute reward score for better explainability.
- Experimental results on the Davis, Kiba, and Allergy datasets to show the feasibility of the proposed framework.

The remaining part of this paper is organized as follows. In the next section, we discuss related work and explain about GNN and MCTS approaches. In Sect. 3, we present the proposed framework. In Sect. 4, we present the experimental results. Section 5 contains a discussion. The last section consists of a summary and conclusions.

2 Related Work and Background

In this section, we explain the related work and present the background approaches.

2.1 Related Work

By extending the notion of frequent pattern mining, an approach to extract frequent subgraphs is proposed in [24]. It also discusses applications in bioinformatics and chemical analysis. The study in [13] introduces a molecule generation approach that automatically discovers frequent principal subgraphs and assembles them into final output molecules using a two-step strategy. The work in [17] provides an overview of subgraph mining approaches.

Research efforts such as [27], DGraphDTA [11], and visGReMLIN [20], are being made by extending deep learning approaches such as convolutional neural networks (CNNs) [36], RNNs [25], GNNs [18,28] to improve the accuracy of drug-target affinity prediction. These models employ methods like frequent subgraph mining, sequence alignment, and contact map prediction, which can be computationally intensive. Computational models like GraphDTA [18] aid in drug repurposing by predicting interaction strengths between new drug-target pairs. Despite its advantages, GraphDTA may struggle with generalization to unseen drug-target pairs significantly different from the training data, especially concerning novel drugs or targets and underrepresented interactions in the dataset.

The work in [21] introduces the Protein-ligand with adversarial augmentations network (PLA-Net), consisting of two modules that extract information from proteins and ligands to develop representations for compatibility analyses. The study in [23] recommends GNN-based DTI prediction by integrating heterogeneous graph information in two steps: constructing a heterogeneous graph based on drug-drug and target-target similarity matrices and the DTI matrix.

A framework named DeepNC is proposed for DTI prediction using GNNs [28]. The complex relationships between drugs and targets are modeled within a graph structure. DeepNC comprises three main components: a graph construction module, a GNN-based feature extraction module, and a prediction module. The study [20] introduces an interactive platform, VisGremlin, to visually explore and interpret these motifs within the context of protein-ligand interfaces.

Deep learning approaches are being extended to address link prediction in semi-bipartite graphs [7]. In another study [1], the performance of 32 network-based machine learning models was evaluated with biomedical datasets. The other work introduces a machine-learning pipeline combining network embedding and gradient-boosted trees for link prediction in identifying novel DTIs [3]. However, methods relying on negative sampling for link prediction can suffer from imbalanced datasets, affecting model performance [38]. Self-explanatory GNNs proposed in [38] aims to balance interpretability and performance, though generating explanations for large-scale link predictions remains computationally intensive for real-time applications.

The opacity of existing deep learning-based models has raised concerns regarding their interpretability and explainability. GNN explainability models, including attention mechanisms [29], subgraph extraction methods [34], post-hoc interpretability techniques [22], and self-interpretable GNNs [8], offer viable approaches to these challenges. Significant research efforts have been directed toward developing methods to explain the predictions of GNNs at different levels: nodes, edges, links, and subgraphs [6,12,16,19,32,35]. GNNExplainer [32] is one of the pioneering methods designed to explain the predictions of GNNs by identifying a compact subgraph and a small subset of node features that are crucial for the prediction. Optimizing for fidelity and sparsity, it provides interpretable explanations for node predictions. In graph-related machine learning tasks, the GNNExplainer is a model-agnostic approach for delivering interpretable explanations for any model-level tasks. The work in [33] introduced a method to interpret GNNs at the model level using reinforcement learning, optimizing graph generation to enhance trained GNNs without ground truth for interpretations of real-world datasets.

A distinctive link prediction framework SEAL [37], built on GNN, it learns based on local surrounding subgraphs, embeddings, and attributes. However, a limited explanation exists, such as the SEAL framework might need to provide more intuitive explanations for their predictions. A general method called ExplaiNE was introduced for explaining link predictions based on network embeddings in terms of the network's existent linkages [12]. Nonetheless, the approach requires careful feature engineering to ensure the network captures relevant graph patterns. Choosing appropriate node features can impact the link prediction performance.

PGExplainer [16] learns a probabilistic graph mask that highlights important edges for the prediction of a given node. It uses a parameterized model to predict the probability of each edge being part of the explanation, thus providing insights into which connections are most influential. GraphSVX [6] extends

Shapley values to GNNs to provide explanations at the subgraph level. By computing the contribution of each subgraph component (nodes and edges) to the prediction, it offers a comprehensive view of important substructures within the graph. XGNN [33] generates graphs that maximize the prediction confidence for a given class, thereby revealing the key structural patterns that the GNN associates with that class.

Unlike existing literature focusing primarily on predicting DTI and enhancing drug candidate selection using various deep-learning models, the proposed framework introduces an explainable GNN-based framework to extract pharmacophores from candidate drugs. By explicitly focusing on pharmacophore extraction, the proposed framework offers pharmacologists a more comprehensive toolset for understanding and leveraging critical structural and chemical features in drug design.

2.2 Background

The proposed framework employs GNN and MCTS approaches. We briefly explain these approaches.

GNN Framework: GNNs have become a powerful framework for leveraging structural information encoded in graph representations [9]. In this context, entities are represented as graphs $G = (V, E)$, where V denotes the set of nodes and E represents the edges between them. Each node v in the graph has an associated feature vector $\mathbf{h}_v$ that encodes specific information about the node. The GNN updates these feature vectors through message-passing mechanisms, where node features are aggregated from their neighbors. This process can be denoted as follows:

$$\mathbf{h}_v^{(k)} = \sigma\left(\sum_{u \in \mathcal{N}(v)} \frac{1}{c_{vu}} \mathbf{W}^{(k)} \mathbf{h}_u^{(k-1)} + \mathbf{W}^{(k)} \mathbf{h}_v^{(k-1)}\right) \tag{1}$$

The update rule for the hidden state $\mathbf{h}_v^{(k)}$ of node v at layer k in a GCN. Here, $\mathcal{N}(v)$ denotes the set of neighboring nodes of v, c_{vu} represents the normalization constant for the edge from node u to node v, $\mathbf{W}^{(k)}$ is the weight matrix for layer k, and σ is the activation function.

Vector embeddings are derived from the final node embeddings after several layers of the GNN. These embeddings aggregate node information into a global representation of the graph. The final vector $\mathbf{e}_G$ is often obtained through pooling operations such as global max pooling (GMP):

$$\mathbf{e}_G = \text{GMP}\left(\{\mathbf{h}_v^{(L)} \mid v \in V\}\right) \tag{2}$$

where $\mathbf{h}_v^{(L)}$ represents the node features after L layers of the GNN. These embeddings capture the overall properties of the graph and can be used in various downstream tasks by feeding them into a prediction model.

About MCTS: MCTS is a heuristic search algorithm widely used in decision-making problems with large and complex search spaces [2]. The algorithm constructs a search tree by performing simulations of potential actions and evaluating their outcomes. A key aspect of MCTS is the Upper Confidence Bound (UCB) criterion, which is mathematically expressed as:

$$\text{UCB}(i) = \bar{X}_i + C\sqrt{\frac{\ln N}{n_i}} \tag{3}$$

where $\bar{X}_i$ is the average reward of node i, N is the total number of simulations conducted from the parent node, n_i is the number of simulations for node i, and C is a constant that balances exploration and exploitation. This equation guides the selection of nodes by favoring those with high potential rewards and those that are less explored, thereby promoting effective exploration of the search space.

To extract important features using MCTS with the UCB approach, the algorithm begins by initializing the search tree with the initial state. During each iteration, MCTS simulates actions such as expanding nodes or selecting edges and updates the search tree based on the outcomes. The UCB criterion is used to choose nodes for expansion by balancing the exploration of new nodes and the exploitation of nodes with known high rewards. This iterative process continues, dynamically adjusting the search strategy based on the observed results. The algorithm focuses on finding the features that most significantly influence the overall representation. These features are prioritized based on their potential to contribute valuable insights into the structure, highlighting the most critical elements.

The MCTS based approach is extended in [35] to extract the potential subgraph from the given graph. We have employed this approach in our framework.

3 Proposed Methodology

We present the proposed approach after explaining the basic idea.

3.1 Basic Idea

As mentioned in the introduction, the issue is to identify a potential pharmacophore (or subgraph) from the identified drug candidate (or molecular graph). First, we need to develop a framework to compute the affinity score for the given drug compound so that potential drug compounds can be identified based on the value of the affinity score. Second, given a target protein and potential drug candidate, a methodology has to be developed to identify the potential subgraph (or pharmacophore) from the drug candidate.

For the first issue, we propose the GNN-based model, named GraphNetDTI as shown in Fig. 1, to identify potential drug candidates by making some updations from DeepNC [28], such as passing arguments and additional features for nodes like chirality, formal charge, and aromaticity. The input is chemical compounds (represented as graphs) and target proteins (represented as embedding vectors) with affinity scores and the output is the trained GNN model.

We consider the DTI model as a regression problem. The predicted affinity $\hat{y}$ is defined as follows:

$$\hat{y} = \sigma(\mathbf{W}_o \mathbf{h}_{\mathbf{DP}} + \mathbf{b}_o) \tag{4}$$

Here, $\hat{y}$ represents the predicted output, $\mathbf{W}_o$ is the weight matrix, $\mathbf{h}_{\mathbf{DP}}$ is the hidden representation (possibly after dropout), and $\mathbf{b}_o$ is the bias term. Where h_{DP} represents the interaction module defined below.

$$\mathbf{h}_{DP} = \sigma(\mathbf{W}_f[\text{Readout}(\{\mathbf{h}_v^{(K)} | v \in V\}); \text{Pooling}(H)] + \mathbf{b}_f) \tag{5}$$

The test molecules with predicted binding affinities exceeding the threshold (ψ) are considered as drug candidates. For the second issue, the trained GNN is leveraged to extract potential subgraphs representing pharmacophores. We propose a framework that takes the drug candidate and the trained GraphNetDTI model as an input and extracts the potential pharmacophores as an output. We employ the MCTS algorithm that enumerates the search space of the important subgraphs of the given drug candidate. We obtain an affinity score from the trained GraphNetDTI model for each subgraph. The subgraph with the maximum affinity score is returned as a result. Overall, the proposed approach consists of the following steps.

- Buiding GraphNetDTI Model
- Extracting potential subgraph from the drug candidate

3.2 Buiding GraphNetDTI Model

The input is the training set of candidate drugs, target protein, and the corresponding affinity scores. The output is the trained model. The steps of this process are as follows.

- Representation of drug candidate and target protein
- Representation of combination of drug candidate and target protein
- Training GraphNetDTI model

We explain these steps. The details are also provided in Algorithm 1.

- **Representation of drug candidates and target protein:**
  Each drug candidate is represented as a graph $G_m(CA, IC)$, where CA is chemical atoms, and IC is ionic & covalent bonds (Lines 3–4). The node feature matrix $\mathbf{X} \in \mathbb{R}^{|V| \times d}$ encodes the properties of the atoms in the molecule. Each row of $\mathbf{X}$ is a feature vector $\mathbf{x}_i$ corresponding to an atom v_i. These features include atom type (e.g., carbon, nitrogen), atom degree, aromaticity, and other relevant chemical properties. The adjacency matrix $\mathbf{A} \in \mathbb{R}^{|V| \times |V|}$ represents the connectivity structure of the molecule. If there is a bond between atom i and atom j, then $\mathbf{A}_{ij} = 1$; otherwise, $\mathbf{A}_{ij} = 0$.

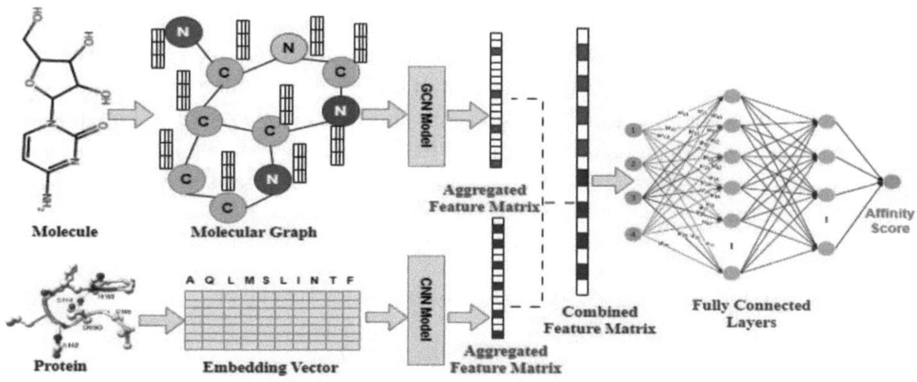

Fig. 1. Proposed GraphNetDTI model architecture

Representation of a Protein (Lines 6–7): Proteins are linear chains of amino acids. Each protein sequence $P = (a_1, a_2, \ldots, a_n)$ is converted into a numerical representation suitable for CNN processing. Each amino acid in the sequence is encoded as a one-hot vector. The one-hot encoded matrix $\mathbf{O} \in \mathbb{R}^{n \times 20}$ has rows corresponding to amino acids, with 20 columns representing the 20 standard amino acids.

- **Representation of combination of candidate drug and protein (Line 18)**
  The combined feature vector is a concatenation of feature vectors of a candidate and protein.
- **Training GraphNetDTI model (Lines 9–16):** The input is a combined feature vector and affinity score. A Graph Convoluton Network (GCN) is used to learn representations of molecular graphs by iteratively updating the node features based on their neighbors' features. The operation of a single GCN layer is given by:

$$\mathbf{H}^{(l+1)} = \sigma\left(\mathbf{D}^{-1/2}\mathbf{A}\mathbf{D}^{-1/2}\mathbf{H}^{(l)}\mathbf{W}^{(l)}\right) \tag{6}$$

Here, $\mathbf{H}^{(l)}$ is the hidden state matrix at layer l, $\mathbf{D}$ is the degree matrix, $\mathbf{W}^{(l)}$ is the weight matrix for layer l, and σ is the activation function (e.g., ReLU). After several GCN layers, the node embeddings are aggregated into a single graph-level representation using a pooling function.

A CNN is used to extract features from the one-hot encoded protein sequences. The CNN operation involves several key layers: The convolutional layer applies multiple filters to the input one-hot encoded sequence to detect local patterns. The operation for the k-th filter is defined as:

$$\mathbf{C}^{(k)} = \sigma\left(\mathbf{W}^{(k)} * \mathbf{E} + \mathbf{b}^{(k)}\right) \tag{7}$$

where the output feature map $\mathbf{C}^{(k)}$ obtained by applying a filter with weights $\mathbf{W}^{(k)}$ to the input one-hot encoded sequence $\mathbf{E}$, followed by adding a bias term $\mathbf{b}^{(k)}$ and applying an activation function σ.

Each filter slides over the input matrix $\mathbf{E}$ and computes dot products between the filter weights and the input sub-matrices, followed by adding the bias and applying the activation function. This results in a feature map $\mathbf{C}^{(k)}$ that captures specific local patterns in the protein sequence. The pooling layer reduces the dimensionality of the feature maps while retaining the most important information (P). After pooling, the feature maps are flattened into a single vector:

$$\mathbf{z}_P = \text{flatten}(\mathbf{P}) \tag{8}$$

This vector is then passed through a fully connected layer:

$$\mathbf{f}_P = \sigma\left(\mathbf{W}_f \mathbf{h}_{\mathbf{DP}} + \mathbf{b}_f\right) \tag{9}$$

where $\mathbf{W}_f \in \mathbb{R}^{m \times p}$ is the weight matrix of the fully connected layer, transforming from m to p dimensions. $\mathbf{b}_f \in \mathbb{R}^p$ is the bias term snd σ is an activation function, typically ReLU. The output $\mathbf{f}_P$ is a fixed-length feature vector representing the protein.

The loss function for the GraphNetDTI model is designed to optimize the prediction of drug-target binding affinities. Let $\mathcal{D} = \{(G_i, P_i, y_i)\}_{i=1}^N$ be the training dataset, where G_i represents the molecular graph of the i-th drug, P_i denotes the protein sequence, and y_i is the true binding affinity. The model leverages GNNs to encode G_i into a feature vector $\mathbf{h}_G$ and CNNs to encode P_i into a feature vector $\mathbf{h}_P$. The predicted binding affinity $\hat{y}_i$ is obtained by combining these vectors through a fully connected layer σ:

$$\hat{y}_i = \sigma(\mathbf{W}[\mathbf{h}_G; \mathbf{h}_P] + \mathbf{b}) \tag{10}$$

Here, $\mathbf{W}$ is the weight matrix, $\mathbf{b}$ is the bias vector, and $[\mathbf{h}_G; \mathbf{h}_P]$ denotes the concatenation of the feature vectors. The loss function $\mathcal{L}$ is defined as the mean squared error (MSE) between the predicted and true binding affinities:

$$\mathcal{L}(\Theta) = \frac{1}{N}\sum_{i=1}^{N}(y_i - \hat{y}_i)^2 \tag{11}$$

where Θ represents the model parameters. To incorporate regularization and prevent overfitting, we add L2 regularization terms:

$$\mathcal{L}_{\text{reg}}(\Theta) = \frac{\lambda}{2}\left(\|\mathbf{W}_{\text{GNN}}\|_F^2 + \|\mathbf{W}_{\text{CNN}}\|_F^2 + \|\mathbf{W}\|_F^2\right) \tag{12}$$

where λ is the regularization parameter and $\|\cdot\|_F$ denotes the Frobenius norm. The total loss function becomes:

$$\mathcal{L}_{\text{total}}(\Theta) = \mathcal{L}(\Theta) + \mathcal{L}_{\text{reg}}(\Theta) \tag{13}$$

This loss function ensures that the model learns to accurately predict binding affinities while maintaining generalization by controlling the complexity of the learned parameters.

Algorithm 1. Drug-Target Interaction Prediction

Input: Chemical Molecules, Protein Sequences, Binding Affinities
Output: Drug Candidates

for each molecular sequence **do**
  Generate molecular graph G_m with atom features $\mathbf{h}_v^{(0)}$
end for
for each protein sequence **do**
  Generate embedding vector $(\mathbf{E})$
end for
for each pair of (G_m) and (P) **do**
  for $l = 0$ to $L - 1$ **do**
    for each node v **do**
      Update node features according to Eq. 1.
    end for
    Aggregate feature matrix for graph G_m:
    $\mathbf{h}_G = \text{GMP}(\{\mathbf{h}_v^{(L)} \mid v \in V_i\})$
    Extract the CNN-based feature vector using Eq. 7
  end for
  Aggregate feature matrix for E from Eq. 2.
  Combine the aggregated features of G_m and E
  Calculate binding score through dense layers $\hat{y}$
end for

3.3 Extracting Potential Subgraph from the Drug Candidate

The issue is to extract the potential subgraph from the drug candidate graph which gives the highest binding affinity score. For this, we extend the MCTS-based framework and propose the model, ExplainbleDeepGNN as shown in Fig. 2. The process involves systematically exploring the subgraphs from the given candidate graph, and selecting the potential subgraph to explore further based on a reward score. First, we consider the entire candidate graph as a root node. The main issue is enumerating subgraphs from the given candidate graph. For this, we sort the nodes of the graphs based on the degree, remove the least degree node, select the largest connected component, and calculate the UCB value. Next, we remove the second least degree node and repeat the same. In this way, we repeat the step at most 15 times and identify the largest connected components. For each connected component or subgraph, we calculate the binding affinity score by the trained GraphNetDTI model. Next, for each subgraph, the reward score and then the UCB value are computed. The reward score is a Gaussian probability density calculated using the predicted binding affinity from the trained GraphNetDTI model and the ground truth data. The subgraph with high UCB is selected and the process is repeated for till each subgraph contains at least three nodes. Among the subgraphs, the subgraph with the highest reward score is selected as a potential subgraph.

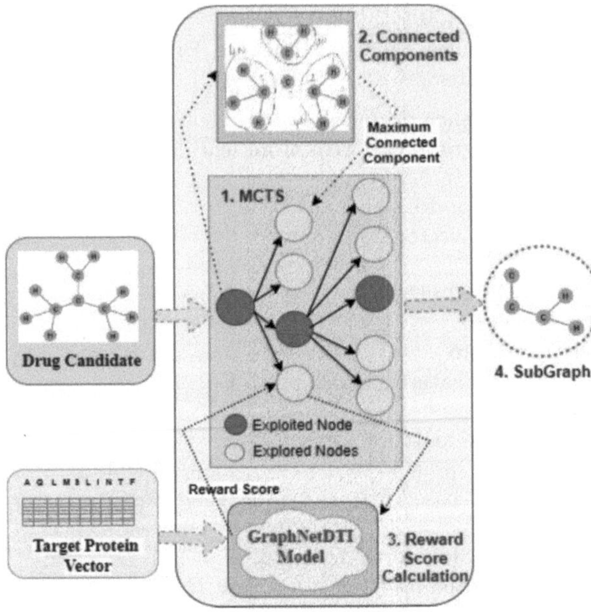

Fig. 2. Proposed ExplainbleDeepGNN model architecture

The input is the candidate graph, G. The proposed MCTS search consists of two steps: Creating a Monte Carlo Search Tree and searching the potential subgraph.

Step 1: Creating a Monte Carlo Search Tree: In this step, we build the MCTS tree for the given candidate graph G by following the following steps.

1.1 Initialize the root node of the tree T with the whole graph G. Initialize the node $< R, V, P >$, where R is the reward value, V is the number of visits and P is the total reward.
1.2 If the number of atoms in G is less than or equal to 3, terminate.
1.3 From G, identify a set L of the largest connected components up to 15, by removing nodes from the current graph ordered by increasing degree. Add elements of L as a child to T.
   1.3.1 If the number of atoms in each element of L is less than or equal to 3, terminate and go to (1.4). Otherwise, compute UCB for each element of L and update the reward score in T for the corresponding node. The graph G', with the highest UCB value, is selected by following Procedure 1.
   1.3.2 From G', identify a set of largest connected components up to 15, by removing nodes from G' ordered by increasing degree and update L. Add elements of L as children to G' in T. Go to step 1.3.1.
1.4 Backtrack the path to the root and update the values of V (number of visits of the node) and P (total reward) for the nodes in the path in T.
1.5 With the updated tree T as input, repeat the steps 1.2 to 1.4, at most 10 times (rollouts).

Step 2: Searching the Potential Subgraph: Iterate over every node in the Monte Carlo Search Tree T and find the node with the highest reward value R.

The procedure to compute the UCB of a subgraph G' from G is given below.

Procedure 1: Compute UCB of a Subgraph G' of G: The steps to calculate UCB of a given subgraph are as follows.

1. For G', the binding affinity is predicted with the trained GraphNetDTI model in Algorithm 1.
2. Reward Score Calculation: Gaussian probability density is calculated with the predicted ($\hat{y}_i$) binding affinity and ground truth value (y_i) and considered as a reward score for the node. Let σ represent the standard deviation of the density.

$$P(\hat{y}_i) = \frac{1}{\sqrt{2\pi\sigma^2}} e^{\left(-\frac{(y-\hat{y}_i)^2}{2\sigma^2}\right)} \tag{14}$$

3. The UCB is calculated by using the reward score P($\hat{y}_i$) (Refer Background Section).

4 Experimental Results

The experiments are conducted on three benchmark datasets: Davis [5], Kiba [26], and Allergy drug datasets [30] to show the effectiveness of the proposed approach in extracting the pharmacophores from candidate drug compounds. The summary of datasets is given in Table 1. The Davis dataset contains 30,056 DTI pairs with K_d values ranging from 5.0 to 10.8 (nM). This dataset provides the most comprehensive and detailed evaluation of the reactivity of known and clinical kinase inhibitors across the entire kinome. The Kiba dataset contains 1,18,255 DTI pairs. In this dataset, the binding affinities are measured as Kiba scores, ranging from 0.0 to 17.2. The Allergy drug dataset contains 372 DTI samples and is curated by collecting drugs for allergic reactions and their targets from DrugBank [31]. Ligands interacting with Allergy drug targets, with K_d values from 1.0 to 15.0 (nM), are obtained from the BindingDB database. Using the RDKit [14] framework, we convert molecules into a Simplified Molecular Input Line Entry System (SMILES) string format. K_d stands for the dissociation constant, which is a measure of the affinity between a drug (ligand) and its target (typically a protein).

Table 1. Summary of benchmark datasets.

Dataset	Number of Pairs	Number of Molecules	Number of Proteins
Davis	30,056	286	35
Kiba	118,255	68	442
Allergy	372	2111	299

The experiments are conducted on the *ADA cluster*, which comprises 42 Boston SYS-7048GR-TR nodes, each equipped with dual Intel Xeon E5-2640 v4 processors and 40 virtual machines per node. ADA's aggregate theoretical peak performance is 47.62 TFLOPS. We utilized 20 virtual machines, each with 2 GB of memory, for our experiments. Python 3.2 is employed for implementation.

Table 2. Hyperparameters and optimal values of the proposed two-phase model.

Hyperparameter	Optimal value	Phase
Batch Size	256	GraphNetDTI
Learning Rate	0.0005	GraphNetDTI
Epochs	1000	GraphNetDTI
Activation Function	ReLU	GraphNetDTI
Dropout	0.2	GraphNetDTI
Number of GCN layers	3	GraphNetDTI
Number of nodes to explore	15	ExplainableDeepGNN
Minimum Atoms	3	ExplainableDeepGNN
Number of Simulations	10	ExplainableDeepGNN
Reward Function	Gaussian probability density	ExplainableDeepGNN

Table 2 presents the hyperparameters and their optimal values used in the proposed two-phase model, which consists of GraphNetDTI and ExplainableDeepGNN. In the GraphNetDTI model, a batch size of 256 was employed to process data in manageable portions, while the learning rate was set to 0.0005 to ensure gradual convergence during training. The model was trained for 1000 epochs, using the ReLU activation function to introduce non-linearity. A dropout rate of 0.2 was applied to prevent overfitting and three GCN layers were utilized to capture intricate molecular relationships. In the ExplainableDeepGNN model, at most 15 nodes (or subgraphs) were explored to focus on the key molecular substructures. The minimum number of atoms considered in the subgraphs was set to 3, and 10 simulations were run to ensure the robustness of the proposed model. A Gaussian probability density function was used as the reward function to guide the extraction of pharmacophores. These hyperparameters were chosen to optimize the performance and explainability of the proposed model.

We employ Mean Absolute Error (MAE), Mean Squared Error (MSE), and Concordance Index (CI) metrics to evaluate performance. MAE measures the average absolute difference between predicted and actual binding affinities, providing insight into prediction accuracy. MSE calculates the average squared difference, highlighting the magnitude of prediction errors. CI assesses the model's ability to rank interactions correctly, reflecting the concordance between predicted and actual outcomes.

The DTI model leverages graph structures to model intricate relationships between entities, such as drugs and targets, effectively capturing the nuanced interactions within the drug-target network. This heightened capability leads to reduced MSE and MAE values, indicative of higher prediction accuracy and narrower CI values, denoting increased confidence in the model's predictions. Thus, the GNN model demonstrates higher predictive prowess in discerning drug-target interactions across various training epochs, as evidenced by the metrics observed on both the Davis and Kiba datasets. For the Davis dataset, the model achieved an MSE of 0.252, an MAE of 0.242, and a CI of 0.871. For the Kiba dataset, the model exhibited error rates with an MSE of 0.141 and an MAE of 0.198 with a higher CI score of 0.875. Eventhough, Allergy dataset has a very less number of samples compared to Devis and Kiba, we extracted the results and achieved an MSE of 12.02, and a CI of 0.58. This is due to higher samples in Kiba. Table 3 highlights the results obtained by state-of-the-art models in the Davis and Kiba datasets. It can be observed that the GNN-DTI model provides the best results overall, with the best MSE values on both datasets (0.242 & 0.141). The GraphNetDTI model matches the best CI values in both datasets (0.871 & 0.875), and thus evaluating both metrics together demonstrates that the GNN-DTI model provides the best results.

Table 3. Comparison results of various models on the Davis and Kiba datasets

Model	Davis		Kiba	
	CI	MSE	CI	MSE
SimBoost [10]	0.869	0.282	0.836	0.222
DeepDTA [39]	0.876	0.262	0.861	0.199
GraphDTA [18]	0.858	0.308	0.871	0.176
DeepGS [15]	0.880	0.252	0.862	0.188
GraphNetDTI	0.871	0.242	0.875	0.141

Table 4 shows the comparison results of binding affinities of the identified drug candidates, ground truth, and extracted pharmacophores. The average binding affinities (ABA) of the Allergy, Davis, and Kiba drug candidates are 4.140, 6.433, and 12.279, respectively. We have calculated the predicted average binding affinities for drug candidates (ABA_{dc}) and molecular subgraphs (ABA_{msg}) extracted from these drug candidates, along with ground truths for the three datasets. Figure 3 shows the ABA values of drug candidates and subgraphs along with ground truth values for Davis, Kiba, and Allergy datasets. The number of drug candidates identified from the Allergy, Davis, and Kiba test datasets are 45, 50, and 113, respectively. We extracted the best molecular subgraph from each drug candidate. From Table 4, it can be noted that ABA_{msg} values are very similar to the ABA_{dc} in the three datasets, varying by only $\pm$ 0.026, $\pm$ 0.349, and $\pm$ 0.737 for Allergy, Davis, and Kiba, respectively. These results demonstrate the effectiveness of the ExplainableDeepGNN model in predicting binding affinities and extracting meaningful subgraphs while maintaining proximity to the ground truth across different datasets.

Table 4. Performance results of ABA values. Here, ABA_{dc} and ABA_{msg} indicate ABA of the drug candidate and the corresponding molecular subgraph, respectively

Dataset	ABA_{dc}	ABA_{msg}	Ground Truth	$\Delta(ABA_{dc}, ABA_{msg})$
Davis	6.433	6.083	6.417	$\pm$ 0.349
Kiba	12.279	11.545	12.427	$\pm$ 0.737
Allergy	4.140	4.114	4.864	$\pm$ 0.026

Figure 4 shows the number of atoms in drug candidates and their corresponding subgraphs, as extracted from the Davis and Kiba dataset. Each graph ID on the x-axis represents a specific drug candidate, while the y-axis indicates the number of atoms. The red bars correspond to the total number of atoms in each drug candidate, whereas the blue bars represent the number of atoms in the extracted subgraphs, which are potential pharmacophores identified using the proposed model. The consistent difference between the red and blue bars highlights the reduction in complexity when focusing on key molecular subcomponents (subgraphs) that are crucial for drug-target interaction, while the original drug candidates generally exhibit a higher number of atoms. The results show how the proposed model isolates essential substructures from larger drug molecules.

Figure 5 illustrates a molecule and its corresponding molecular graph, along with the subgraph extracted using the ExplainableDeepGNN model from three different datasets. The molecular graph representation encapsulates the intricate connectivity of atoms and bonds within the molecule, effectively translating chemical structures into graph data. The ExplainableDeepGNN model identifies crucial substructures by focusing on key nodes and edges, thus isolating subgraphs that are most relevant for prediction tasks. This extraction process highlights the model's ability to pinpoint and leverage specific molecular subcomponents that contribute significantly to its learning outcomes. By analyzing the subgraphs, the model can infer chemical properties and biological activities, offering a deeper understanding of the molecular mechanisms.

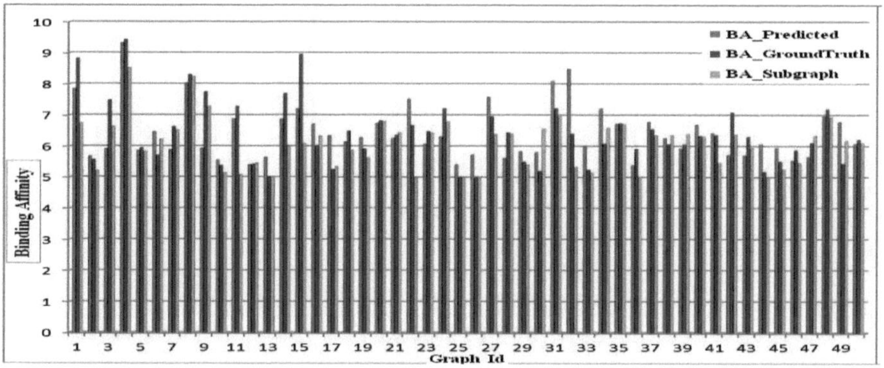

(a) Davis Dataset

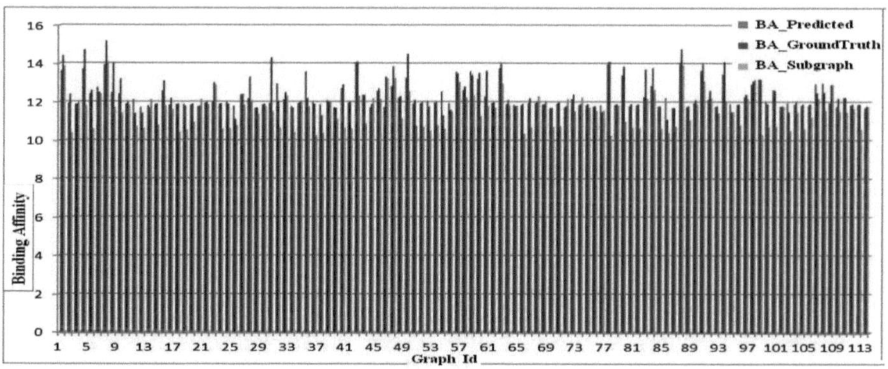

(b) Kiba Dataset

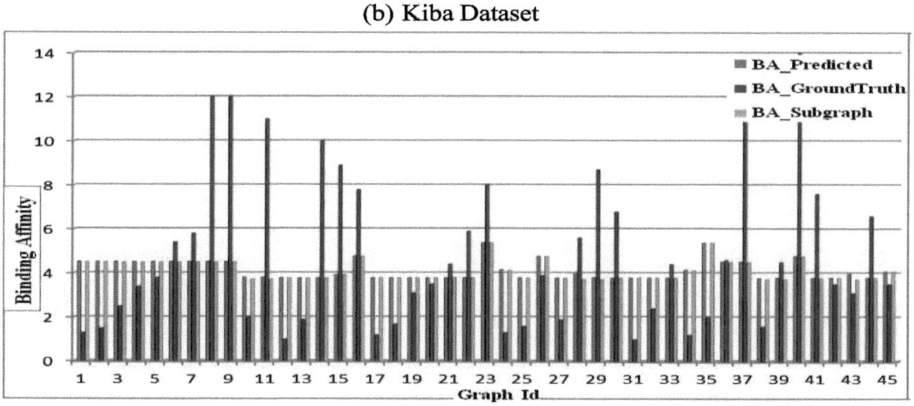

(c) Allergy Dataset

Fig. 3. Binding affinities of drug candidate, ground truth, extracted subgraph.

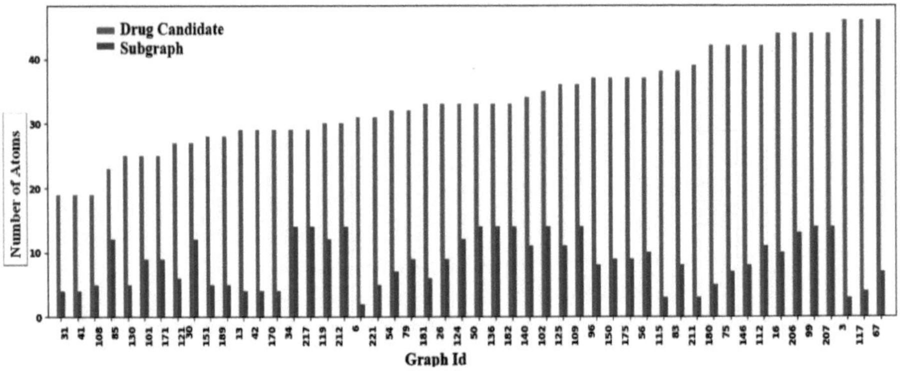

(a) Davis Dataset

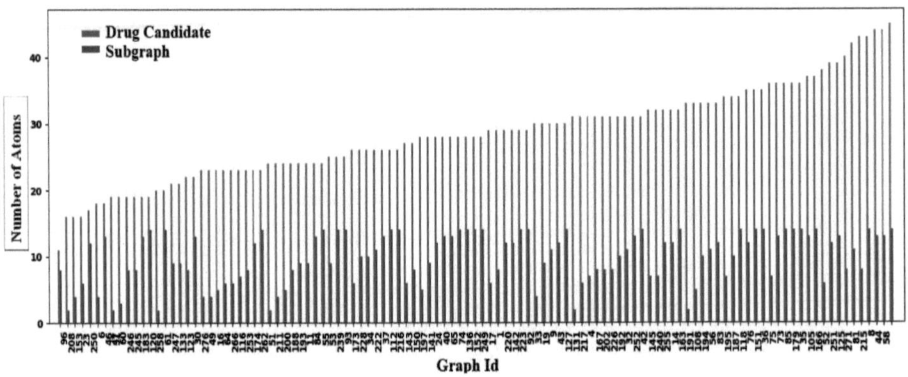

(b) Kiba Dataset

Fig. 4. The number of atoms in drug candidate and subgraph

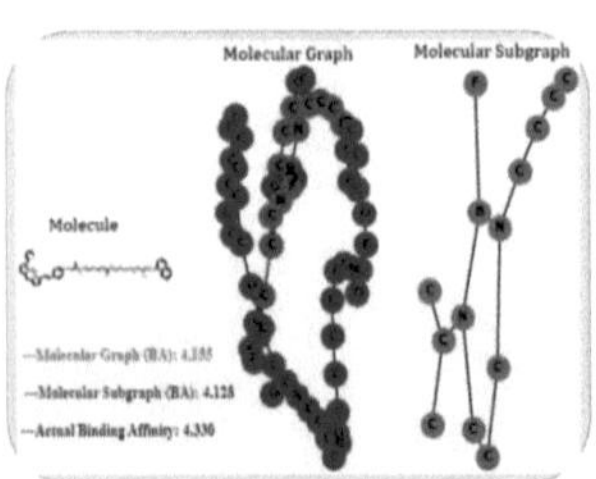

(a) Allergy Dataset

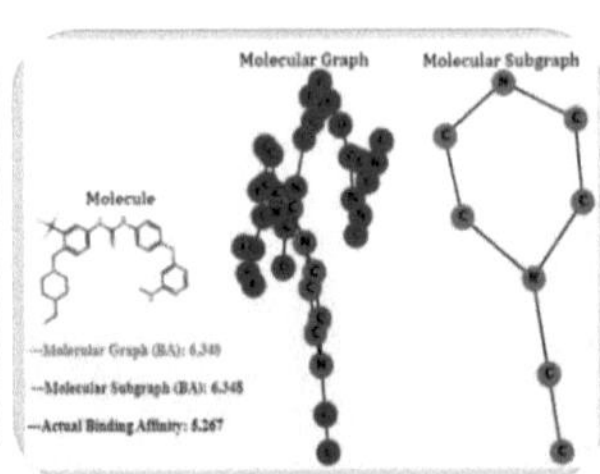

(b) Davis Dataset

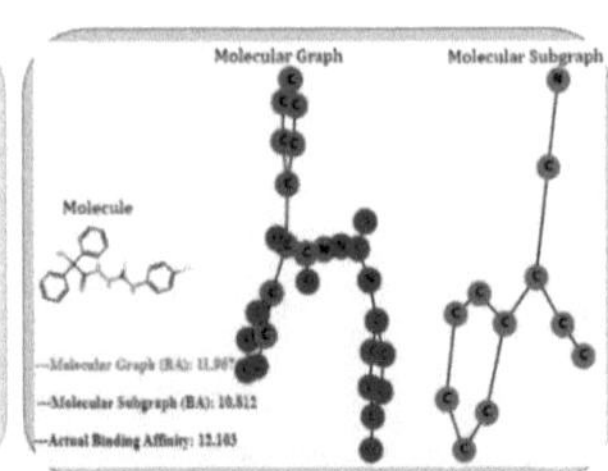

(c) Kiba Dataset

Fig. 5. Extracted subgraphs from three sample drug candidates

5 Discussion

We present discussion and limitations.

- **Extracting multiple pharmacophores from the drug candidate:** In our proposed approach, we focus on extracting one subgraph, or pharmacophore, as the output rather than considering multiple subgraphs for several reasons. First, selecting a single subgraph offers a more interpretable and simplified explanation of the critical features of the drug candidate. Pharmacophores are defined as the essential structural and chemical features necessary for biological interaction, and thus focusing on one dominant subgraph allows us to directly capture the key elements driving DTI. Multiple subgraphs could introduce ambiguity, making it difficult to pinpoint which features are most crucial. Additionally, selecting one subgraph aligns well with the explainability goal of our model, providing a clearer insight into the mechanism behind the drug's effectiveness. However, it is important to acknowledge that drugs may exhibit multi-faceted interactions, and exploring multiple subgraphs in future work could enhance model complexity and performance. Nevertheless, such an approach would require sophisticated selection mechanisms to rank subgraphs according to their significance in interactions.
- **Modeling protein with graph:** Pharmacophores are critical in drug discovery as they define the essential chemical features required for a drug to interact effectively with specific biological targets, guiding the design of new therapeutic agents. GCNs play a pivotal role in DTIs by analyzing complex molecular graphs, which allows for precise modeling of the intricate relationships between drugs and their targets. This capability is particularly important for understanding drug pockets in proteins—specific sites where drugs bind, influencing their effectiveness and specificity. The role of explainability in drug discovery is paramount, as it ensures that models deliver not only accurate predictions but also transparent insights into the mechanisms of drug interactions.

Limitations: While our proposed GraphNetDTI and ExplainableDeepGNN show promising results, some limitations need to be addressed. First, our reliance on high-quality, labeled data for training and validation poses a significant challenge, as the availability of comprehensive datasets can impact model performance. Although our model exhibits strong performance on the Davis and Kiba datasets, its generalizability to other datasets and broader applications needs further validation. Another crucial limitation is that since we introduce explainability through the ExplainableDeepGNN model, comparing its interpretability with state-of-the-art models is difficult due to the need for standardized benchmarks and metrics for explainability. This makes it challenging to assess and validate the effectiveness of our explainability approach quantitatively. Additionally, while the ExplainableDeepGNN model provides valuable insights into molecular interactions, interpreting these insights requires significant domain-specific knowledge, limiting the model's accessibility to non-experts. Addressing these limitations in future work.

6 Conclusion and Future Work

In the process of drug discovery, identifying Drug-Target Interactions (DTIs) is an essential step. In the literature, research efforts are being made to extend machine learning and graph neural networks to identify potential drug candidates. From a drug discovery point of view, a pharmacophore, a subgraph of a drug candidate, is an essential structural and chemical feature necessary for interacting with a specific biological target. The problem of extracting the knowledge of potential pharmacophores from candidate drugs remains largely unexplored. In this paper, we have proposed a two-phase GNN-based framework for identifying drug candidates and extracting potential pharmacophores from these candidates. In this framework, we have developed a GNN-based model to compute the affinity score for the given drug compound. We have also proposed a Monte Carlo Tree Search (MCTS) based algorithm, which leverages the trained GNN to extract potential subgraphs representing pharmacophores. The extensive experimental results show that the three popular benchmark datasets show high predictive accuracy. Furthermore, the proposed model successfully extracts pharmacophores of relatively small size from the candidate drugs extracted from the three datasets. The results show that the proposed framework not only enhances pharmacophore extraction but also enriches the overall interpretability and insight into critical molecular interactions, thus paving the way for more informed drug discovery processes.

As a part of future work, we are going to conduct a validation study of the extracted drug candidates and the corresponding pharmacophores with chemists. We are planning to extend our models to encompass even larger and more diverse datasets, enhancing their generalizability. Extending the proposed framework to identify protein pockets will enhance the model's ability to identify key binding sites more accurately. Exploring the application of these models in personalized medicine and drug repurposing is another exciting direction.

Acknowledgements. This research is supported by the Science and Engineering Research Board (SERB) (Sanction order File No. SRG/2023/000941), Department of Science and Technology (DST), Government of India. We also acknowledge the support of iHub-Data, IIIT Hyderabad, Telangana, India.

References

1. Abbas, K., et al.: Application of network link prediction in drug discovery. BMC Bioinform. **22**, 1–21 (2021)
2. Abramson, B.: The Expected-Outcome Model of Two-Player Games. Morgan Kaufmann (2014)
3. Amiri Souri, E., Laddach, R., Karagiannis, S.N., Papageorgiou, L.G., Tsoka, S.: Novel drug-target interactions via link prediction and network embedding. BMC Bioinform. **23**(1), 121 (2022)
4. Chaslot, G.M.J.-B.C.: Monte-Carlo tree search (2010)

5. Davis, M.I., et al.: Comprehensive analysis of kinase inhibitor selectivity. Nat. Biotechnol. **29**(11), 1046–1051 (2011)
6. Duval, A., Malliaros, F.D.: GraphSVX: shapley value explanations for graph neural networks. In: Machine Learning and Knowledge Discovery in Databases. Research Track: European Conference, ECML PKDD 2021. Springer (2021)
7. Manoochehri, H.E., Nourani, M.: Drug-target interaction prediction using semi-bipartite graph model and deep learning. BMC Bioinform. **21**, 1–16 (2020)
8. Fang, L., et al.: SIG: Efficient self-interpretable graph neural network for continuous-time dynamic graphs. arXiv preprint arXiv:2405.19062 (2024)
9. Gori, M., Monfardini, G., Scarselli, F.: A new model for learning in graph domains. In: Proceedings of the 2005 IEEE International Joint Conference on Neural Networks, vol. 2, pp. 729–734 (2005)
10. He, T., Heidemeyer, M., Ban, F., Cherkasov, A., Ester, M.: SimBoost: a read-across approach for predicting drug-target binding affinities using gradient boosting machines. J. Cheminform. **9**, 1–14 (2017)
11. Jiang, M., Li, Z., Zhang, S., Wang, S., Wang, X., Yuan, Q., Wei, Z.: Drug-target affinity prediction using graph neural network and contact maps. RSC Adv. **10**(35), 20701–20712 (2020)
12. Kang, B., Lijffijt, J., De Bie, T.: Explaine: an approach for explaining network embedding-based link predictions. arXiv preprint arXiv:1904.12694 (2019)
13. Kong, X., Huang, W., Tan, Z., Liu, Y.: Molecule generation by principal subgraph mining and assembling (2022)
14. Landrum, G., et al.: RDKit: a software suite for cheminformatics, computational chemistry, and predictive modeling. Greg Landrum **8**(31.10), 5281 (2013)
15. Lin, X., Zhao, K., Xiao, T., Quan, Z., Wang, Z.-J., Yu, P.S.: DeepGS: deep representation learning of graphs and sequences for drug-target binding affinity prediction. In: ECAI 2020, pp. 1301–1308. IOS Press (2020)
16. Luo, D., et al.: Parameterized explainer for graph neural network. Adv. Neural. Inf. Process. Syst. **33**, 19620–19631 (2020)
17. Mrzic, A., et al.: Grasping frequent subgraph mining for bioinformatics applications. BioData Min. **11** (2018)
18. Nguyen, T., Le, H., Quinn, T.P., Nguyen, T., Le, T.D., Venkatesh, S.: GraphDTA: Predicting drug-target binding affinity with graph neural networks. Bioinformatics **37**(8), 1140–1147 (2021)
19. Ribeiro, M.T., Singh, S., Guestrin, C.: "Why should i trust you?" Explaining the predictions of any classifier. In: Proceedings of the 22nd ACM SIGKDD International Conference on Knowledge Discovery and Data Mining, pp. 1135–1144 (2016)
20. Ribeiro, V.S., et al.: visGReMLIN: graph mining-based detection and visualization of conserved motifs at 3D protein-ligand interface at the atomic level. BMC Bioinform. **21**(2), 1–12 (2020)
21. Puentes, P.R., et al.: Predicting target-ligand interactions with graph convolutional networks for interpretable pharmaceutical discovery. Sci. Rep. **12**(1), 1–17 (2022)
22. Schlichtkrull, M.S., De Cao, N., Titov, I.: Interpreting graph neural networks for NLP with differentiable edge masking. arXiv preprint arXiv:2010.00577 (2020)
23. Shao, K., Zhang, Y., Wen, Y., Zhang, Z., He, S., Bo, X.: DTI-HETA: prediction of drug-target interactions based on GCN and gat on heterogeneous graph. Briefings in Bioinformatics **23**(3) (2022)
24. Sharma, A., Reinher, S., Mehta, D., Wu, B.: FLEXIS: FLEXible frequent subgraph mining using maximal independent sets (2024)

25. Sivangi, K.B., Amilpur, S., Dasari, C.M.: ReGen-DTI: a novel generative drug target interaction model for predicting potential drug candidates against SARS-COV2. Comput. Biol. Chem. **106**, 107927 (2023)
26. Tang, J., et al.: Making sense of large-scale kinase inhibitor bioactivity data sets: a comparative and integrative analysis. J. Chem. Inf. Model. **54**(3), 735–743 (2014). PMID: 24521231
27. Torng, W., Altman, R.B.: Graph convolutional neural networks for predicting drug-target interactions. J. Chem. Inf. Model. **59**(10), 4131–4149 (2019)
28. Tran, H.N.T., Thomas, J.J., Malim, N.H.A.H.: DeepNC: a framework for drug-target interaction prediction with graph neural networks. PeerJ **10**, e13163 (2022)
29. Veličković, P., Cucurull, G., Casanova, A., Romero, A., Lio, P., Bengio, Y.: Graph attention networks. arXiv preprint arXiv:1710.10903 (2017)
30. Wang, Y., et al.: PubChem bioassay: 2017 update. Nucleic Acids Res. **45**(D1), D955–D963 (2017)
31. Wishart, D.S., et al.: DrugBank: a knowledgebase for drugs, drug actions and drug targets. Nucleic Acids Res. **36**(suppl_1), D901–D906 (2008)
32. Ying, Z., Bourgeois, D., You, J., Zitnik, M., Leskovec, J.: GNNExplainer: generating explanations for graph neural networks. Adv. Neural Inf. Process. Syst. **32** (2019)
33. Yuan, H., Tang, J., Hu, X., Ji, S.: XGNN: towards model-level explanations of graph neural networks. In: Proceedings of the 26th ACM SIGKDD International Conference on Knowledge Discovery & Data Mining, pp. 430–438 (2020)
34. Yuan, H., Haiyang, Yu., Gui, S., Ji, S.: Explainability in graph neural networks: a taxonomic survey. IEEE Trans. Pattern Anal. Mach. Intell. **45**(5), 5782–5799 (2022)
35. Yuan, H., Yu, H., Wang, J., Li, K., Ji, S.: On explainability of graph neural networks via subgraph explorations. In: International Conference on Machine Learning, pp. 12241–12252. PMLR (2021)
36. Zhang, H, Liu, X., Cheng, W., Wang, T., Chen, Y.: Prediction of drug-target binding affinity based on deep learning models. Comput. Biol. Med. 108435 (2024)
37. Zhang, M., Chen, Y.: Link prediction based on graph neural networks. Adv. Neural Inf. Process. Syst. **31** (2018)
38. Zhu, H., Luo, D., Tang, X., Xu, J., Liu, H., Wang, S.: Self-explainable graph neural networks for link prediction. arXiv preprint arXiv:2305.12578 (2023)
39. Öztürk, H., Özgür, A., Ozkirimli, E.: DeepDTA: deep drug-target binding affinity prediction. Bioinformatics **34**(17), i821–i829 (2018)

GDAdaP: A Domain Adaptation Framework for Gene Dependency Prediction in Lung Cancer

Da Tan and Carson K. Leung(✉)

Department of Computer Science, University of Manitoba, Winnipeg, MB, Canada
Carson.Leung@UManitoba.ca

Abstract. Gene dependency (GD) prediction has become a critical tool in cancer research, aiding in the identification of therapeutic targets and providing insights into tumor biology. In this paper, we present a domain adaptation framework for predicting GD scores across lung cancer subtypes (e.g., adenocarcinoma, mesothelioma, squamous cell carcinoma) using CRISPR-Cas9 and ribonucleic acid (RNA) interference data adapted to The Cancer Genome Atlas (TCGA) patient tumor data. Our framework leverages a deep learning model to transfer knowledge from cell line data (source domain) to patient tumor data (target domain), enhancing generalization across different environmental settings for cancer cells. The model is evaluated using metrics such as mean absolute error (MAE), Pearson correlation, and R^2, demonstrating high accuracy and robustness on both source and target domains. These findings underscore the value of GD scores in capturing molecular distinctions across lung cancer subtypes and inform strategies for personalized treatment.

Keywords: bioinformatics · computational biology · gene dependency · prediction · data science · machine learning · domain adaptation · clustered regularly interspaced short palindromic repeats and its associated protein 9 (CRISPR-Cas9) · The Cancer Genome Atlas (TCGA) · lung cancer

1 Introduction

Nowadays, big data are everywhere. With advancements in technology, huge amounts of data can be easily generated and collected at a rapid rate from a wide variety of real-life application domains. Embedded in these big data are implicit, previously unknown but useful information. Data science solutions [1,2]—which make good use of data mining [3–7], machine learning [8,9] (including deep learning [10,11]), mathematical and statistical techniques [12,13], and visualization [14–16]—help discover the useful information that was previously hidden in the big data. These solutions can be applicable for application areas like health informatics [17–22], bioinformatics and computational biology [23–29], climatology [30], finance [31,32], sports analytics [33], and transportation analytics [34]. Prediction [21–26,30–34] is a common task in many application areas. In

T. Aldwairi et al. (Eds.): BICOB 2025, CCIS 2535, pp. 163–176, 2025.
https://doi.org/10.1007/978-3-031-94039-2_13

this paper, we focus on prediction in bioinformatics and computational biology. Specifically, we present a bioinformatics and computational biology framework for gene dependency (GD) prediction.

GD refers to the extent to which a cell relies on the function of a particular gene for survival or proliferation. In cancer cells, GD is highly genetic context dependent [35], and understanding it is critical for revealing potential therapeutic targets. If a gene is essential for the survival of cancer cells but not for normal cells, inhibiting it could selectively kill cancer cells, leading to new treatments [36]. GD analysis helps in identifying "driver" genes responsible for cancer progression and can guide precision medicine by predicting which genes to target in specific cancer types.

GD predictions based on high-throughput genomic data offer critical insights into understanding cancer biology and identifying potential therapeutic targets. GD data can be derived from various experimental methods [37] such as:

- a genome editing technology—called CRISPR-Cas9—that uses an enzyme to modify deoxyribonucleic acid (DNA) for silencing the gene at the DNA level (i.e., "knockout"), where (a) CRISPR refers to clustered regularly interspaced short palindromic repeats and (b) Cas9 refers to CRISPR-associated protein 9; or
- a cellular technology—called RNA interference (RNAi) screen—that uses the gene's own DNA sequence to reduce gene expression at the messenger ribonucleic acid (mRNA) level (i.e., "knockdown").

These methods have been increasingly used to predict cancer-specific vulnerabilities by associating gene knockouts with tumor phenotypes.

Cancer is a complex disease with diverse molecular subtypes, each characterized by unique genetic alterations and dependency profiles. Understanding these profiles across different cancer subtypes is key to identifying subtype-specific therapeutic targets. Traditional methods often struggle to account for the large, multi-dimensional nature of the data, particularly when generalizing findings from one cancer data source to another [38]. Domain adaptation approaches [39] have emerged as powerful tools for transferring knowledge from one data domain to another, particularly when training data for the target domain is scarce or heterogeneous.

The *motivation* for this study is to bridge the gap between different domains—especially from experimental cell line settings to patient tumor samples—to predict GD scores for cancers. This research specifically focuses on the prediction of GD scores for lung cancer subtypes using domain adaptation techniques. These lung cancer subtypes[1] include:

- adenocarcinoma, which usually starts in the mucus-producing cells in the outer parts of the lung;
- squamous cell carcinoma, which often starts in the flat cells that line the bronchi near the middle of the lungs; and
- mesothelioma, which develops in thin tissue—called mesothelium—in the lining of the lungs.

[1] https://www.cancercenter.com/cancer-types/lung-cancer/types.

We focus on lung cancer because the training dataset encompassing all cancer types can be exceedingly large, posing computational challenges, and GD-based targeted therapies remain underexplored for lung cancer subtypes. We incorporate data from the following sources:

- Broad CRISPR-Cas9 and RNAi datasets [40]; and
- The Cancer Genome Atlas (TCGA) datasets [41].

By doing so, we aim to provide a more accurate picture of the molecular dependencies driving the subtypes of lung cancers, potentially uncovering new targets for therapeutic intervention.

Our key contributions of this paper include our domain adaptation framework, called GDAdaP, to predict GD score on target domain tumor cell data. It is a non-trivial application of a novel deep learning model that optimizes GD score predictions for lung cancer. We employ domain adaptation techniques to account for the shifts in data distributions between different cancer datasets, thus improving the generalizability of the model. We aim to enhance the utility of GD predictions in clinical decision-making, helping to identify more precise and personalized cancer treatment strategies.

The remainder of this paper is organized as follows. The next section discusses related work. Section 3 describes our domain adaptation framework GDAdaP for predict GD score on target domain tumor cell data. Then, Sect. 4 shows evaluation results. Finally, Sect. 5 draws the conclusions.

2 Related Work

GD is essential in understanding cancer vulnerabilities and therapeutic strategies. Several studies have investigated the prediction of gene dependencies using high-throughput genomic data, with a focus on various machine learning and deep learning approaches. For example, as a large-scale initiative, DepMap [40] provides GD data based on CRISPR-Cas9 [42] and RNAi [43] screens, which have been utilized extensively in cancer genomics studies [44]. These efforts have been fundamental in uncovering critical dependencies in cancer—including both universal and cancer-specific genes—paving the way for targeted drug development [45].

Adaptation techniques [46,47] have gained increasing attention in the field of cancer research, particularly due to the challenge of transferring models trained on experimental dataset (e.g., cancer cell lines) to real-world clinical data (e.g., patient tumor samples). Domain adaptation approaches [39] aim to mitigate distributional shift between training and target domains, enabling models to generalize better when faced with different experimental conditions or data sources. For instance, Zhang et al. [48] utilized domain adaptation techniques to transfer knowledge from one domain to another, yielding positive results. Similar models [49] have been shown to effectively capture complex dependencies in high-dimensional cancer genomics data.

While numerous studies have explored for cancer vulnerability, few have specifically focused on GD differences in lung cancer subtypes. Recent advancements in domain adaptation methods offer the potential to generalize GD predictions across different lung cancer subtypes, which can provide more personalized treatment strategies. Our approach applies a deep learning model with domain adaptation to predict GD scores for adenocarcinoma, mesothelioma, and squamous cell carcinoma, offering novel insights into the molecular dependencies driving these subtypes.

3 Our Domain Adaptation Framework for GD Prediction

3.1 Overview

In this study, we tackle the challenge of domain adaptation for GD prediction. Domain shift and out-of-distribution (OOD) issues are prevalent in healthcare-related fields due to variations in environmental and experimental contexts [50]. For instance, in cancer genetics, it is well-recognized that preclinical and human tumor domains experience OOD challenges [51,52]. The preclinical domain, encompassing cell line data and patient-derived xenografts (PDX) in mice, is influenced by specific laboratory factors. In contrast, the human tumor domain is shaped by complexities such as the immune system and tumor micro-environment.

In our specific problem, illustrated in Fig. 1, we address the domain adaptation task involving GD scores derived from CRISPR-Cas9 and RNAi cell line data, alongside unlabeled TCGA patient tumor data. The objective is to bridge the gap between cell line and tumor domains, leveraging domain adaptation techniques to predict GD scores for TCGA tumors. The source domain comprises labeled RNAi data, while the target domains include labeled CRISPR-Cas9 data and unlabeled TCGA tumor data.

To achieve robust prediction in the face of these domain discrepancies, we employ maximum mean discrepancy (MMD) loss [53] to align the feature distributions between source and target domains because its advantages in handling high-dimensional data. By minimizing MMD loss, the proposed framework effectively transfers learned representations, enabling accurate GD score predictions for the target domain.

GDAdaP is a deep learning framework designed to integrate and process two distinct data modalities: gene expression ($\mathbf{X}_{\text{exp}}$) and gene fingerprint ($\mathbf{X}_{\text{fp}}$) data. As illustrated in Fig. 2, training begins by utilizing deep neural networks on three datasets: CRISPR-Cas9, RNAi, and TCGA. Mean squared error (MSE) losses are computed for both CRISPR-Cas9 and RNAi datasets to optimize predictions, while MMD losses are calculated to align feature distributions between the RNAi and TCGA domains, as well as between the CRISPR-Cas9 and TCGA domains.

As depicted in Fig. 3, the framework employs three specialized subnetworks, each dedicated to extracting feature representations from its respective data modality. These extracted features are subsequently concatenated to form a

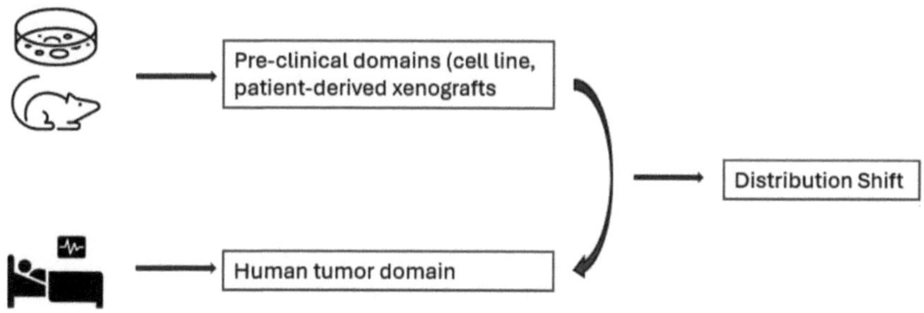

Fig. 1. Goal: To bridge the gap between preclinical cell line data and patient tumor data for accurate GD predictions.

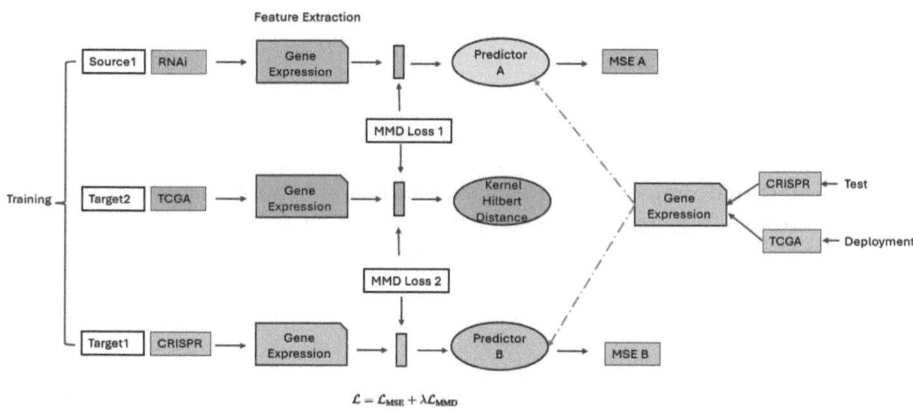

Fig. 2. Our GDAdaP framework is trained on RNAi and CRISPR-Cas9 data to adapt and predict GD scores for the TCGA domain.

shared representation, which is then passed through fully connected layers to predict the dependency score y. This modular design allows GDAdaP to effectively learn information from both data types. More specifically, GDAdaP consists of three main components:

1. A feature encoder neural network for dimensionality reduction and representation learning from gene expression data;
2. An encoder neural network to extract functional fingerprints for each GD; and
3. A final neural network that concatenates gene expression features with fingerprints to predict GD scores.

Formally, the framework can be expressed as:

$$\hat{y} = g\left([f_{\text{exp}}(\mathbf{X}_{\text{exp}}) \mid\mid f_{\text{fp}}(\mathbf{X}_{\text{fp}})]\right) \tag{1}$$

where f_{exp} and f_{fp} are the feature extractors (neural networks shown in Fig. 3 for gene expression and fingerprint data, respectively, and $g(\cdot)$ represents the fully connected layers for prediction.

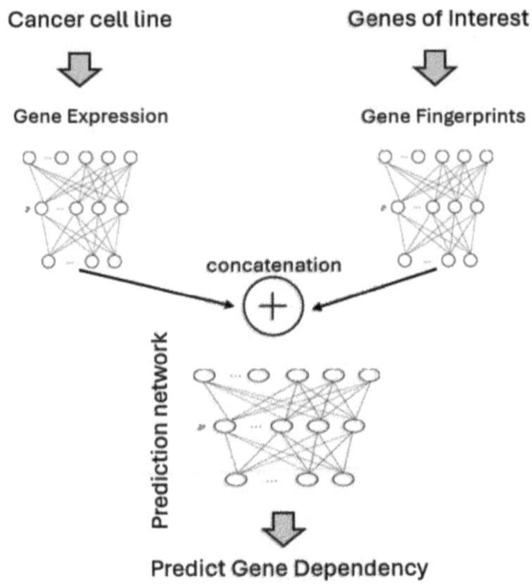

Fig. 3. Our GDAdaP framework consists of three neural network components: one for extracting features from gene expression data, another for processing gene fingerprint data, and a final network that combines these features to generate predictions.

3.2 Loss Functions

We optimize the model using a combination of supervised regression loss on the labeled domain and distribution alignment losses on the target domain to bridge the domain gap.

The **supervised loss** is the MSE between the predicted dependency scores $\hat{y}$ and the true scores y in the source domain:

$$\mathcal{L}_{\text{MSE}} = \frac{1}{N_s} \sum_{i=1}^{N_s} (\hat{y}_i - y_i)^2 \tag{2}$$

where N_s is the number of samples in the source domain.

To align the feature distributions between the source and target domains, we employ the **maximum mean discrepancy (MMD) loss**. The MMD computes the distance between the mean embeddings of two distributions in a reproducing kernel Hilbert space (RKHS) [53]. Given source features $\mathbf{H}_s$ and target features $\mathbf{H}_t$, the MMD is defined as:

$$\begin{aligned} \text{MMD}^2(\mathbf{H}_s, \mathbf{H}_t) = \frac{1}{N_s^2} \sum_{i=1}^{N_s} \sum_{j=1}^{N_s} k(\mathbf{h}_i^s, \mathbf{h}_j^s) + \frac{1}{N_t^2} \sum_{i=1}^{N_t} \sum_{j=1}^{N_t} k(\mathbf{h}_i^t, \mathbf{h}_j^t) \\ - \frac{2}{N_s N_t} \sum_{i=1}^{N_s} \sum_{j=1}^{N_t} k(\mathbf{h}_i^s, \mathbf{h}_j^t) \end{aligned} \tag{3}$$

where $k(\cdot,\cdot)$ is a kernel function (e.g., RBF or linear), and N_t is the number of target samples. With Eq. (3), the total MMD loss can be computed as the sum of MMD loss at each of the three levels of feature representations—namely, the final shared features, gene expression features, and fingerprint features:

$$\mathcal{L}_{\text{MMD}} = \text{MMD}^2(\mathbf{H}_s, \mathbf{H}_t) + \text{MMD}^2(\mathbf{H}_{\text{exp},s}, \mathbf{H}_{\text{exp},t}) + \text{MMD}^2(\mathbf{H}_{\text{fp},s}, \mathbf{H}_{\text{fp},t}) \quad (4)$$

where $\mathbf{H}_{\text{exp}}$ and $\mathbf{H}_{\text{fp}}$ are the intermediate feature representations for gene expression and fingerprint data, respectively.

By combining Eqs. (2) and (4), the **total loss** for training can be computed as a weighted sum of the MSE loss on the source domain and the MMD loss for aligning distributions:

$$\mathcal{L} = \mathcal{L}_{\text{MSE}} + \lambda\, \mathcal{L}_{\text{MMD}} \quad (5)$$

where λ is a hyperparameter controlling the contribution of the MMD loss. We set $\lambda = 1$ in this study.

3.3 Training Procedure

The training process involves the following steps:

1. Source Domain Supervised Training: Minimize $\mathcal{L}_{\text{MSE}}$ to train the model on labeled RNAi data.
2. Target Domain Adaptation (CRISPR-Cas9): Align the feature distributions of the RNAi source domain and CRISPR-Cas9 target domain using $\mathcal{L}_{\text{MMD}}$.
3. Unsupervised Adaptation (TCGA): Further align RNAi source domain features with those of the unlabeled TCGA data using $\mathcal{L}_{\text{MMD}}$.
4. Compute $\mathcal{L}_{\text{MSE}}$ on CRISPR-Cas9, and $\mathcal{L}_{\text{MMD}}$ on both CRISPR-Cas9 and TCGA for model selection.

For each training epoch, the model alternates between these steps, with updates performed using backpropagation and gradient descent.

4 Evaluation

4.1 Setup

We presented GDAdaP to predict GD for patient tumor samples. GDAdaP was first trained on representations derived from high-dimensional gene expression profiles of cell line samples and subsequently adapted to the target domain of patient tumor samples, as shown in Fig. 3. To evaluate GDAdaP's domain adaptation capabilities, three GD Cancer Cell Line (CCL) datasets from different sources were utilized:

- 628 CCLs with RNAi screening data [54] served as the source domain.
- 1058 CCLs with Broad CRISPR-Cas9 screening data (Broad DepMap 2024 Q2 [40]) constituted the target and supervised domains to which the model needed to adapt.

– 1100 TCGA lung cancer samples [41] constituted the target but unsupervised domains to which the model needed to adapt.

Among them, gene expression datasets for RNAi and CRISPR-Cas9 experiments were downloaded from the Broad DepMap Portal [40]. CRISPR-Cas9 data were selected as a target domain because it has been shown to be more distributionally similar to TCGA data [55]. The TCGA dataset was obtained from the Treehouse Tumor Compendium[2] filtered for lung cancer samples based on meta-clinical information. This resulted in 1100 TCGA lung cancer samples used for predictions.

Details on the train, validation, and test splits are summarized in Table 1. Note that, since the MSE loss is not computed on the target domain datasets (CRISPR-Cas9 and TCGA), we did not create validation sets for these datasets.

Table 1. Experiment design

Data	#CCLs				Is supervised?
	Training	Validation	Test	Total	
RNAi (source domain) [54]	502	63	63	628	✓
CRISPR-Cas9 (target domain) [40]	846	–	212	1058	✓
TCGA (target domain) [41]	880	–	220	1100	×

Gene fingerprints were represented as binary vectors indicating the involvement of each GD in 3494 molecular signatures from the Molecular Signature Database (MSigDB) [56]. In this representation, a value of 1 denotes involvement, while 0 indicates non-involvement. These fingerprints provide a functional profile relevant to molecular processes within the cell and act as unique representations of each GD. On average, a GD is involved in 38.5 molecular signatures, following a tailed distribution, with a significant portion having more than 40 involvements.

GD scores, which estimate gene effects, were derived using CERES [57], with more negative values indicating higher dependency. We selected 1130 genes for dependency prediction, guided by prior studies [37]. These genes were chosen by intersecting CRISPR-Cas9 and RNAi datasets, followed by filtering for standard deviation SD ≥ 0.2 across the two datasets. Table 1 details the sample sizes of the source and target domain datasets, along with their respective train/validation/test splits.

[2] Tumor Compendium v10 Public PolyA (July 2019): https://treehousegenomics.soe.ucsc.edu/public-data/previous-compendia.html#tumor_v10_polyA.

4.2 Framework Performance

Here, we present the performance evaluation of our GDAdaP framework. Specifically, we conduct a *clustering analysis* on the predicted GD scores for the target domain TCGA dataset to assess how well the framework captures the underlying structure of different cancer subtypes.

Our framework was evaluated on test sets from both the source and target domains. As this is a regression task, performance was assessed using metrics such as:

- mean absolute error (MAE),
- R^2, and
- Pearson's correlation coefficient ρ.

On the source domain test set, the model demonstrated high accuracy with $MAE = 0.212$, $R^2 = 0.876$ and $\rho = 0.760$, as shown in Fig. 4. Similarly, the model adapted well to the target domain, achieving $MAE = 0.213$, $R^2 = 0.869$ and $\rho = 0.755$ on the target domain test set. These results highlight the model's robust performance across domains, with comparable accuracy between the source and target domains.

4.3 Baseline Comparison

To benchmark GDAdaP against existing methods, three alternative approaches were tested using the same train/validation/test split:

- DeepDEP [37], which employs a deep learning framework that pretrains the model on the target TCGA domain using unsupervised techniques but does not address domain shift.
- Least absolute shrinkage and selection operator (LASSO) regression
- Principal component analysis followed by a ridge regression model (PCA + Ridge).

Table 2 presents a detailed comparison of all baseline models. It is evident that GDAdaP outperformed the alternatives significantly, particularly for the R^2 metric. GDAdaP achieved an $R^2 = 0.869$, whereas DeepDEP achieved $R^2 = 0.506$. The LASSO and PCA+Ridge models performed considerably worse, with R^2 values of 0.393 and −0.003, respectively. This substantial improvement underscores the effectiveness of our domain adaptation strategy in capturing the underlying GD patterns.

Figure 5 illustrates the loss trajectory throughout the training process. We tracked the RNAi training and validation MSE losses, the CRISPR-Cas9 MSE and MMD losses for training and test sets, and the TCGA MMD loss. Additionally, the trajectories for Pearson's correlation and R^2 on the target domain validation set are reported. As shown in the figure, all loss metrics exhibit a clear downward trend, while Pearson's correlation and R^2 values consistently improve during training. We selected the model at epoch 40 as the optimal checkpoint, as the validation losses stabilize and reach a plateau at this point.

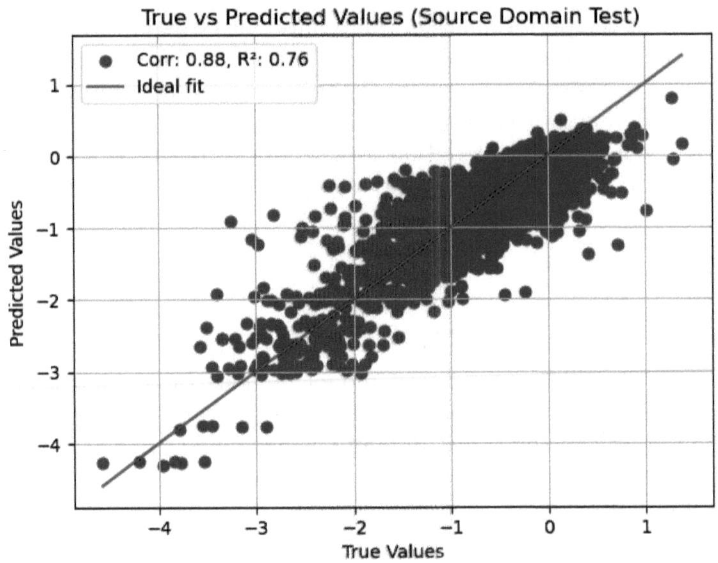

(a) Predicted GD scores compared to true values for the source domain (RNAi) test dataset.

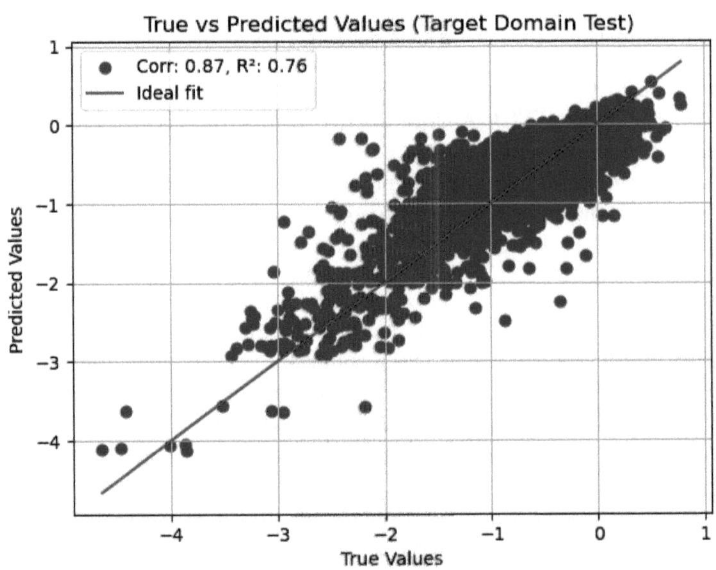

(b) Predicted GD scores compared to true values for the target domain (CRISPR-Cas9) dataset.

Fig. 4. Prediction for source and target domain datasets.

Table 2. Baseline model comparison

Data Set	MAE	R^2	Pearson correlation
DeepDEP	0.301	0.506	0.725
LASSO	0.334	0.393	0.649
PCA+Ridge	0.429	−0.003	0.018
Our source domain	**0.212**	**0.876**	**0.760**
Our target domain	**0.213**	**0.869**	**0.755**

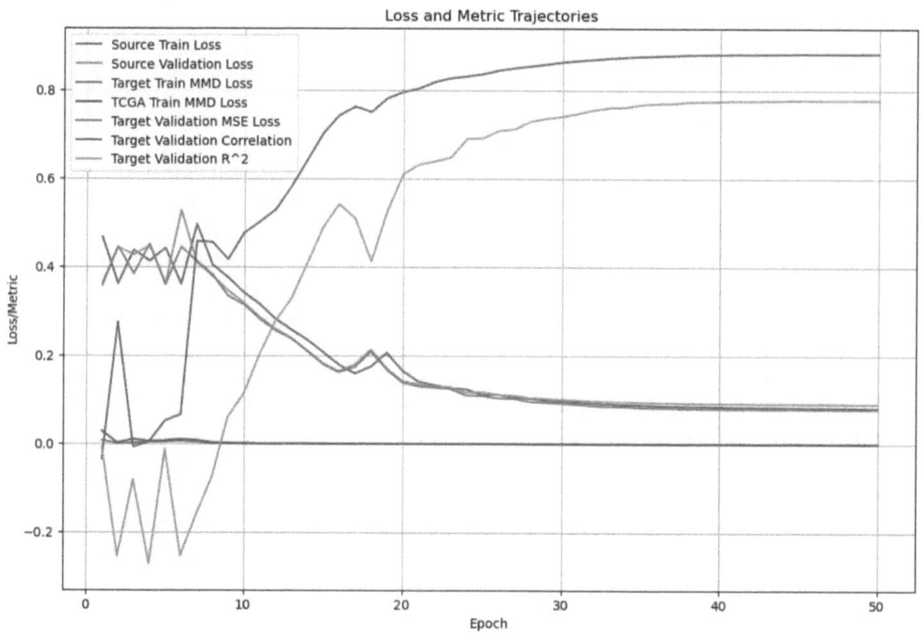

Fig. 5. Trajectories of key metrics during the training process.

5 Conclusions

In this study, we developed a novel domain adaptation deep learning framework for predicting gene dependency (GD) scores across different data domains. Using RNAi data as the source domain, and CRISPR-Cas9 cell line and TCGA tumor samples as the target domains, we demonstrated the ability of the model to accurately predict GD scores for target domain data, capturing intricate molecular dependencies in lung cancer subtypes. The model achieved competitive performance on standard regression metrics, surpassing baseline methods such as LASSO regression, PCA+Ridge, and DeepDEP, thereby highlighting the robustness of our domain adaptation strategy. Overall, this study underscores the potential of domain-adaptive approaches in integrating disparate datasets to uncover actionable biological insights. By bridging the gap between RNAi, CRISPR-Cas9 and TCGA data, our model offers a scalable framework for identifying gene dependencies and therapeutic targets in cancer research. *Ongoing and future directions* include: (a) conducting additional analyses (e.g., t-SNE clustering analysis, KEGG pathway analysis, survival analysis), (b) extending this framework to additional cancer types, and (c) exploring its application in predicting synthetic lethality and drug response to further enhance precision oncology efforts.

Acknowledgements. This work is partially supported by Natural Sciences and Engineering Research Council of Canada (NSERC) and University of Manitoba.

References

1. Avagyan, G., et al.: An environmental data science solution for data analytics exploration of traffic interaction on air quality. In: IEEE COMPSAC 2024, pp. 1806–1811 (2024)
2. Mai, T.H.D., et al.: A data science solution to integrate weather data for energy consumption analysis. In: IEEE IRI 2024, pp. 69–75 (2024)
3. Lakshmanan, L., et al.: The segment support map: scalable mining of frequent itemsets. ACM SIGKDD Explor. **2**(2), 21–27 (2000)
4. Leung, C.K., Hayduk, Y.: Mining frequent patterns from uncertain data with MapReduce for big data analytics. In: DASFAA 2013, Part I. LNCS, vol. 7825, pp. 440–455 (2013)
5. Rizvee, R.A., et al.: A tree-based framework to mine top-k closed sequential patterns. Appl. Intell. **55**(3), 221:1–221:29 (2025)
6. Roy, K.K., et al.: Mining sequential patterns in uncertain databases using hierarchical index structure. In: PAKDD 2021, Part II. LNCS (LNAI), vol. 12713, pp. 29–41 (2021)
7. Seo, S.H., Salem, S.: Mining approximate frequent dense modules from multiple gene expression datasets. In: BICOB 2020, pp. 129–138 (2020)
8. Alam, M.T., et al.: Supervised graph embedding for classification using discriminating frequent patterns. Int. J. Data Sci. Anal. (JDSA) **19**(2), 337–349 (2025)
9. Romero, R.R., Guevara, E.: Classification of functional near infra red signals with machine learning for prediction of epilepsy. In: BICOB 2020, pp. 41–48 (2020)
10. Leung, C.K., et al.: AI-based sensor information fusion for supporting deep supervised learning. Sensors **19**(6), 1345:1–1345:12 (2019)
11. Zhou, X., et al.: Deep learning-empowered big data analytics in biomedical applications and digital healthcare. IEEE/ACM Trans. Comput. Biol. Bioinf. **21**(4), 516–520 (2024)
12. Lac, L., et al.: Computational frameworks integrating deep learning and statistical models in mining multimodal omics data. J. Biomed. Inform. **152**, 104629:1-104629:13 (2024)
13. Leung, C.K., Singh, S.P.: A mathematical model for friend discovery from dynamic social graphs. In: IEEE/ACM ASONAM 2021, pp. 569–576 (2021)
14. Barkwell, K.E., et al.: Big data visualisation and visual analytics for music data mining. In: IV 2018, pp. 235–240 (2018)

15. Berrios, D., et al.: Visualizing omics data from spaceflight samples using the NASA GeneLab Platform. In: BICOB 2020, pp. 89–98 (2020)
16. Chen, D., et al.: A visual analytics solution for analyzing and mining infectious disease data. In: IV 2024, pp. 297–300 (2024)
17. Ciftci, B., et al.: Disability-based clusters and their clinicodemographic determinants in patients with multiple sclerosis: preliminary results of an unsupervised clustering approach. Mult. Scler. J. **29**(3S), 451–452 (2023)
18. Hryhoruk, C.C.J., Leung, C.K.: Mining sequential patterns with timelines from digital health data. In: IEEE ICDH 2023, pp. 292–294 (2023)
19. Leung, C.K., Nguyen, H.H.: A novel deep learning approach for breast cancer detection on screening mammography. In: IEEE BIBE 2023, pp. 277–284 (2023)
20. Vattheuer, C.N., et al.: Multiple multi-modal methods of malignant mammogram classification. In: IEEE ICHI 2024, pp. 57–66 (2024)
21. Camp, J., Al-Mubaid, H.: Simple evolutionary algorithm for quantifying how medical history factors predict disease outcomes. In: BICOB 2022, pp. 43–52 (2022)
22. Dotzlaw, K.I., et al.: Data analytics and prediction of long COVID cases with fuzzy logic. In: FUZZ-IEEE 2023 (2023)
23. Almalki, B., et al.: Transmembrane protein inter-helical residue contacts prediction using transductive support vector machines. In: BICOB 2023, pp. 35–46 (2023)
24. Almalki, B., Liao, L.: TMHC-MSAT: accurate prediction of inter-helical residue contacts in transmembrane proteins using MSA transformer. In: BICOB 2024, pp. 1–10 (2024)
25. Fung, D., et al.: A self-knowledge distillation-driven CNN-LSTM model for predicting disease outcomes using longitudinal microbiome data. Bioinform. Adv. **3**(1), vbad059:1–vbad059:10 (2023)
26. Sun, Y., et al.: iNGNN-DTI: prediction of drug-target interaction with interpretable nested graph neural network and pretrained molecule models. Bioinformatics **40**(3), btae135:1–btae135:12 (2024)
27. Leung, C.K.: Biomedical informatics: state of the art, challenges, and opportunities. BioMedInformatics **4**(1), 89–97 (2024)
28. Lin, E., et al.: scGMM-VGAE: a Gaussian mixture model-based variational graph autoencoder algorithm for clustering single-cell RNA-seq data. Mach. Learn.: Sci. Technol. **4**(3), 035013:1–035013:12 (2023)
29. Tan, D., et al.: A guided variational autoencoder for targeted molecule optimization in drug discovery. J. Healthc. Inform. Res. (JHIR) (2025). https://doi.org/10.1007/s41666-025-00189-6
30. Bian, F., et al.: A digital twin system to classify tornado climatology data for predicting future disasters. In: IEEE DSAA 2024, pp. 461–470 (2024)
31. Camara, R.C., et al.: Fuzzy logic-based data analytics on predicting the effect of hurricanes on the stock market. In: FUZZ-IEEE 2018, pp. 576–583 (2018)
32. Sababipour Asl, G., et al.: Deep learning-based credit score prediction: hybrid LSTM-GRU model. In: IEEE SSCI 2023, pp. 395–400 (2023)
33. Anuraj, A., et al.: Sports data mining for cricket match prediction. In: AINA 2023, vol. 3. LNNS, vol. 655, pp. 668–680 (2023)
34. Lanting, S.J., et al.: A transportation analytic solution for predicting flight cancellations. In: IEEE IRI 2023, pp. 248–253 (2023)
35. Lin, A., Sheltzer, J.M.: Discovering and validating cancer genetic dependencies: approaches and pitfalls. Nat. Rev. Genet. **21**(11), 671–682 (2020)
36. Arafeh, R., et al.: The present and future of the cancer dependency map. Nat. Rev. Cancer **25**, 59–73 (2025)

37. Chiu, Y., et al.: Predicting and characterizing a cancer dependency map of tumors with deep learning. Sci. Adv. **7**(34), eabh1275 (2021)
38. Peres da Silva, R., et al.: TUGDA: task uncertainty guided domain adaptation for robust generalization of cancer drug response prediction from in vitro to in vivo settings. Bioinformatics **37**(Supplement 1), i76–i83 (2021)
39. Ben-David, S., et al.: A theory of learning from different domains. Mach. Learn. **79**, 151–175 (2010)
40. Tsherniak, A., et al.: Defining a cancer dependency map. Cell **170**(3), 564–576 (2017)
41. Cancer Genome Atlas Research Network: Comprehensive and integrative genomic characterization of human cancers. Cell **169**(7), 1327–1341.e23 (2018)
42. Jinek, M., et al.: A programmable dual-RNA–guided DNA endonuclease in adaptive bacterial immunity. Science **337**(6096), 816–821 (2012)
43. Paddison, P.J., et al.: A resource for large-scale RNA-interference-based screens in mammals. Nature **428**(6981), 427–431 (2004)
44. Itzhacky, N., Sharan, R.: Prediction of cancer dependencies from expression data using deep learning. Mol. Omics **17**(1), 66–71 (2021)
45. Behan, F.M., et al.: Prioritization of cancer therapeutic targets using CRISPR–Cas9 screens. Nature **568**(7753), 511–516 (2019)
46. Jiang, J., et al.: Tumor-aware, adversarial domain adaptation from CT to MRI for lung cancer segmentation. In: MICCAI 2018, Part II, pp. 777–785 (2018)
47. Abbet, C., et al.: Self-rule to multi-adapt: Generalized multi-source feature learning using unsupervised domain adaptation for colorectal cancer tissue detection. Med. Image Anal. **79**, 102473 (2022)
48. Zhang, Y., et al.: Bridging theory and algorithm for domain adaptation. In: ICML 2019, pp. 7404–7413 (2019)
49. Chen, M.M., et al.: Predicting cancer cell line dependencies from the protein expression data of reverse-phase protein arrays. JCO Clin. Cancer Inform. **4**, 357–366 (2020)
50. Fittall, M.W., van Loo, P.: Translating insights into tumor evolution to clinical practice: promises and challenges. Genome Med. **11**(1), 20 (2019)
51. Hou, X., et al.: Opportunities and challenges of patient-derived models in cancer research: patient-derived xenografts, patient-derived organoid and patient-derived cells. World J. Surg. Oncol. **20**(1), 37 (2022)
52. Picca, F., et al.: From cancer to immune organoids: Innovative preclinical models to dissect the crosstalk between cancer cells and the tumor microenvironment. Int. J. Mol. Sci. **25**(19), 10823 (2024)
53. Gretton, A., et al.: A kernel method for the two-sample-problem. In: NIPS 2006, pp. 513–520 (2006)
54. McFarland, J.M., et al.: Improved estimation of cancer dependencies from large-scale RNAI screens using model-based normalization and data integration. Nat. Commun. **9**(1), 4610 (2018)
55. Shi, X., et al.: TCGA$_{\text{DEPMAP}}$—mapping translational dependencies and synthetic lethalities within the cancer genome atlas. bioRxiv (2022). https://doi.org/10.1101/2022.03.24.485544v2
56. Liberzon, A., et al.: Molecular signatures database (MSigDB) 3.0. Bioinformatics **27**(12), 1739–1740 (2011)
57. Meyers, R.M., et al.: Computational correction of copy number effect improves specificity of CRISPR–Cas9 essentiality screens in cancer cells. Nat. Genet. **49**(12), 1779–1784 (2017)

Combining Non-Negative Matrix Factorization with Molecular Energy Landscape Analysis for Structure Quality Estimation of Proteins

Kazi Lutful Kabir(✉)

George Mason University, Fairfax, VA 22030, USA
kkabir@gmu.edu

Abstract. Determining the intrinsic organization of the molecular structure space is crucial in various computational analyses of proteins. Identification of the inherent structural groups plays a vital role in this regard. The grouping of tertiary structures of a protein has broad applicability that includes rendering structure-energy landscape, capturing stable and quasi- stable structural states, modeling structural dynamics, and relating them to biological function. Its applicability can be extended further by utilizing the structural groups as the main ingredients to develop methods for single structure selection as well as for structure quality assessment. Exploring along the direction, this paper puts forward a novel method that aggregates structure proximity matrix factorization with structure-energy landscape analysis. Even though the method has several utilities, this article focuses on demonstrating its effectiveness in addressing the structure quality assessment problem.

Keywords: Energy Landscape · Matrix Factorization · Protein Tertiary Structure · Structure Quality Estimation

1 Introduction

The 3D arrangement of amino acids in a protein, known as its tertiary structure, is largely responsible for determining the protein's function in living cells [11]. AlphaFold 3 has made significant progress in determining these structures, successfully positioning all amino acids in human proteins [1]. However, these approaches (AlphaFold 2 [22, 40] and AlphaFold 3) overlook the flexibility of protein structures i.e., their ability to adopt multiple conformations that are functionally relevant [34]. To address this, some researchers have developed methods to explore the protein structure-energy landscape [3, 25] whereas some others have relied on physics-based [13, 31] insights, revealing transitions between structures and organizing them into groups. These groups capture stable and semi-stable states that support different cellular activities of proteins. Furthermore, some methods use level sets [33] to extract these states, while some others apply graph partitioning to identify structural groups [23, 27], also known as macrostates.

Over the years, researchers have developed approaches dedicated for grouping structures and modeling protein dynamics. These methods have rarely been focused on the

T. Aldwairi et al. (Eds.): BICOB 2025, CCIS 2535, pp. 177–188, 2025.
https://doi.org/10.1007/978-3-031-94039-2_15

structure quality assessment (QA) task considering it as a completely separate problem. However, there are a few methods [5, 24] that covered both of these directions i.e., productive in terms of identifying structural groups and reasonable in evaluating the quality of the structures. In fact, this combined perspective is an active research direction [18] in the field of computational biology.

With a view to exploring this direction, a novel method is being proposed here that combines structure similarity matrix factorization with structure-energy landscape analysis to group structures as well as evaluate the quality of the structures. The method offers a variety of utilities, including functioning as a non- parametric consensus clustering tool, assigning scores to individual structures for single-structure selection, grouping structures for multi-structure selection, and combining group assignments to identify outlier structures, etc. However, the focus of this paper is to demonstrate the effectiveness of this method for structure quality assessment.

The rest of the paper is organized as follows. A summary of related works is provided in Sect. 2. The methodology is described in Sect. 3, and the results of the experimentation and evaluation are related in Sect. 4. The paper concludes with a summary and pointer of future work in Sect. 5.

2 Related Works

Initially, QA methods were focused on single-structure selection. These methods assigned a single score per structure and relied heavily on molecular energy functions. The scope of these methods was limited even after including statistical terms [19]. Multi-structure methods took a different route by clustering/grouping structures based on similarity [43] instead of scoring them. They adopted different strategies to identify the best group/cluster as well as to select the top structure of the identified best group/cluster. In fact, conventional clustering approaches performed poorly in these settings [2, 23]. Over time, multi-model methods improved by incorporating energy [4] alongside the structural information. They remained dominant for a while [37] until larger structure databases paved the way for single-structure ML approaches. Early ML methods leveraged shallow models to score structures [36], while more recent deep neural network-based methods [9, 16, 20] are now capable of generating very accurate scores to be considered as benchmarks [17].

Even though explored via other applications of computational biology such as uncovering mutational signatures in human cancer [7], grouping genes for molecular pattern discovery [14]; it remained quite unexplored in this domain for some time until NMF-MAD method in [5] utilized it to decompose a matrix formed with energy-based features [6]. Then, SNMF-DS [24] eliminated the need for features by decomposing a structure similarity matrix in a non-parametric setting utilizing the eigen-gap heuristic [42] to automatically determine the optimal number of groups. SNMF-DS outperformed NMF-MAD, MUFOLD-CL [43], and single- structure methods, such as SBROD [28]. Work in [26] shifted from matrix to tensor factorization by developing a hybrid method that can serve as both single and multiple-structure selection framework. Factorization-based methods have been effective due to the capability of handling sparse and extremely imbalanced datasets. Besides, continuous improvements in the quality of the computed structure also played a supportive role in this regard.

3 Methodology

The proposed method is referred to as NMF-EL QA (Non-negative Matrix Factorization with Energy Landscape for Quality Assessment) which can be decomposed into four phases. Phase I organizes the given structures into groups via symmetric non-negative matrix factorization. Phase II captures the multi-view of the structure data and then utilizes the same to prune some structures to come up with a reduced set. Phase III takes the reduced set of structures and then groups them in three different ways: the mechanism for the first two are same as that of phase I (but with different similarity matrices) and the other additionally considers the potential energy of the structures (along with the structural coordinate information) to extract the energy basins in the energy landscape. Phase III ends by aggregating those three group assignment outcomes into one. Phase IV employs a structure weighting scheme to assign a score for each structure, representing the quality of the structure within the set. A schematic is related in Fig. 1.

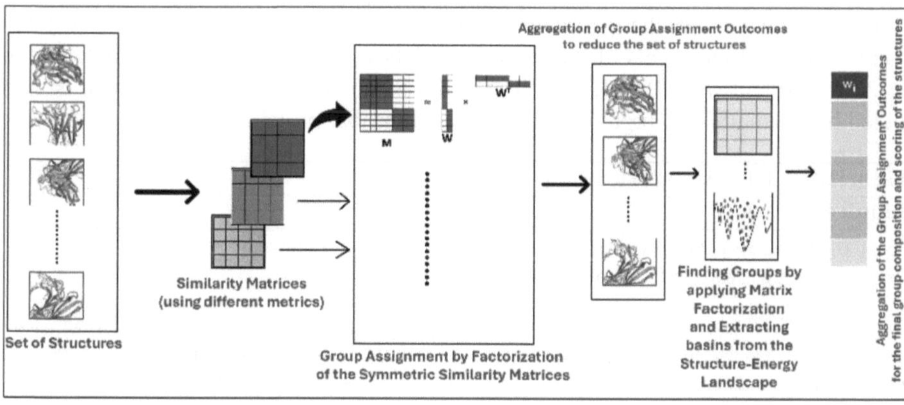

Fig. 1. A high-level schematic of the proposed method NMF-EL QA

3.1 Phase I: From Structures to Structural Groups

From Structures to Structure Similarity Matrix: The process begins with the Cartesian coordinates of the n computed structures of a given protein target as input to create a similarity matrix where each of those represents a tertiary structure generated via a template-free approach using the Rosetta AbInitio protocol [32] with PyRosetta. To minimize computational complexity while constructing the similarity matrix M (of size $n \times n$); each tertiary structure is simplified by retaining only the main-chain carbon (CA) atoms, discarding side-chain and other backbone atoms. Each entry of the matrix, $M_{i,j}$ denotes the similarity between two structures at positions i, j in a set of n structures.

$$M_{i,j} = \begin{cases} \frac{1}{RMSD(i,j)} & \text{if } i \neq j \\ 1 & \text{if } i = j \end{cases}$$

It is to be noted that there are different metrics for comparing two structures which capture different aspects and often provide complementary information [38]. Hence, matrix M can be constructed in different ways using any popular metrics like RMSD, TM-Score, GDT-TS, MaxSub score, etc.

From Similarity Matrix to Number of Structural Groups: Symmetric NMF is employed on the matrix to find the number of structural groups in a non-parametric manner as described in [24] using EigenGap heuristic [42]. The pairwise structure similarity matrix M is used to search for the b nearest neighbors of each structure. Thus, the procedure formulates a nearest-neighbor graph (nngraph), where structures are vertices connected via edges to their b nearest neighbors. Instead of explicitly constructing this graph, it creates an adjacency matrix A (entries represent whether corresponding structure pairs are connected or not) and a degree matrix D (diagonal matrix capturing the degree of each vertex). These matrices are leveraged to form the Laplacian matrix which undergoes further processing to determine the optimal number of groups (k) for organizing the structures. Algorithm 1 [39] shows the steps for applying EigenGap heuristic to find k. Figure 2 shows the EigenGap curve computed as described over the structures of one of the protein targets of the evaluation dataset. The highest peak is obtained at $k = 23$ for protein with PDB ID 6MSP; i.e., 23 is the number of groups which is then utilized by the method to organize the structures for that protein target.

Algorithm 1 EigenGap Heuristic

Require: M, b

$A \leftarrow$ Adjacency Matrix from $M, b = \sqrt{n}$
$D \leftarrow$ Degree Matrix from A
$L \leftarrow D - A$ //Laplacian Matrix
$L_M \leftarrow D^{-1/2} L D^{-1/2}$ //Symmetric normalized Laplacian Matrix
$\{U, V\} \leftarrow$ Eigen-decompose(L_M) //$U \Leftarrow$ eigenvalues, $V \Leftarrow$ eigenvectors
$u \leftarrow$ diagonal(U) //extract eigenvalues
sort the elements of u in ascending order
for $k \leftarrow 1 \ldots$ *u.length* **do**
  EigenGap(k) $\leftarrow u(k+1) - u(k)$
end for
return k that corresponds to the highest peak of the EigenGap(k)-curve

Organizing the Structures into Groups: The phase continues by applying symmetric NMF to approximate the similarity matrix M using a lower rank factorization WW^T where, the matrix W represents the membership of clusters, identifying the specific groups for the structures. In fact, symmetric NMF [30] solves the following equation for the cluster membership indicator matrix $W \in \mathbb{R}_+^{n \times k}$; where similarity matrix $M \in \mathbb{R}_+^{n \times n}$ (M is a symmetric matrix i.e., $M = M^T$), and $\|\cdot\|_F$ indicates the *Frobenius norm*-based minimization.

$$min_{W \geq 0} f(W) = \left\| M - WW^T \right\|_F^2 \quad (1)$$

Typically, $k << n$. To solve the specified optimization problem for W, alternating non-negative least squares (ANLS) optimization (with block principal pivoting) is applied

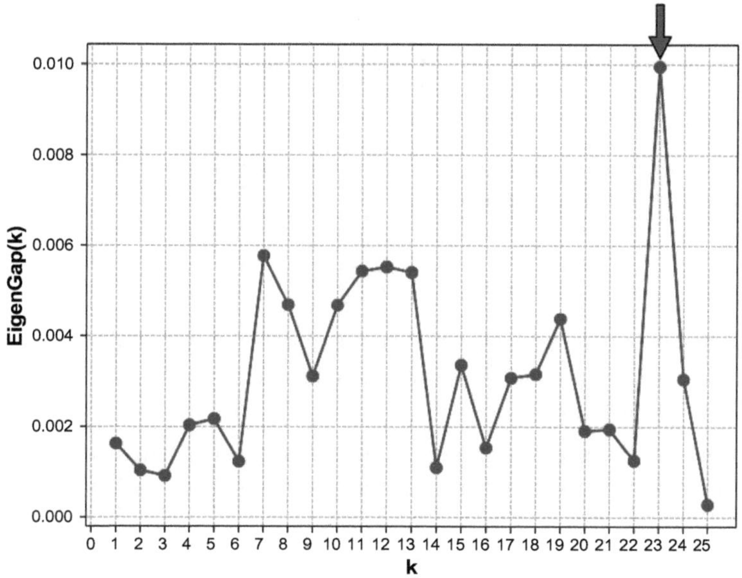

Fig. 2. The EigenGap curve computed over structures of Protein with PDB ID: 6MSP. The highest peak is obtained at $k = 23$ (indicated by an arrow), thus identifying 23 groups. The curve is only shown for the first 25 components (out of 200) in the interest of clarity.

that converges to stationary points [29]. For better convergence, nonnegative double singular value decomposition (NNDSVD) [12] is used to initialize the factor W. The cluster assignments are determined by the largest value in each row of the W matrix.

3.2 Phase II: Structure Pruning

To capture multi-view of the same structure data, Phase I is repeated twice using different similarity metrics. Initially, group assignments are derived from a similarity matrix based on RMSD, as outlined in Phase I. In the subsequent step, Phase I is performed again using a similarity matrix constructed with TM-Score. Finally, the process is repeated with a similarity matrix built on pairwise GDT-TS values for the corresponding structures. Now, there are 3 outcomes for the group assignment of the structures corresponding to the 3 cases mentioned. These outcomes are utilized to reduce the set of structures adopting the mechanism described in [21].

From Group Assignment Labels to Group Assignment Resemblance Matrix: Let's consider that the set of structures for each protein target can be denoted by $S = \{s_1, s_2, s_3, \ldots\ldots\ldots, s_n\}$. A group assignment outcome of S can be represented as a label vector $O \in N^n$ where O_i is the group-label of s_i. Let $G = \{O^{(1)}, O^{(2)}, \ldots\ldots\ldots, O^{(C)}\}$ indicates a set of C group assignment outcomes (here, $C = 3$) of the same set of structures S and each $O^{(i)}$ is a group-label vector of $\left\{O_1^{(i)}, O_2^{(i)}, \ldots\ldots\ldots, O_n^{(i)}\right\}$. Then, the group-label vector is converted into group assignment resemblance matrix, $R^{(i)}$ such that:

$$R_{j,k}^{(i)} = \begin{cases} 1 \text{ if } O_j^{(i)} = O_k^{(i)} \\ 0 \text{ else} \end{cases}$$

Similarly, after converting all of the C group assignment outcomes, an average resemblance matrix can be computed as

$$R' = \frac{\sum_{i=1}^{C} R^{(i)}}{C} \tag{2}$$

It is to be noted that the value of R'_{ij} represents the likelihood that the i-th and the j-th structures are assigned to the identical group and a large value indicates a high probability of the same [21].

Quantization of Group Assignment Uncertainty and Structure Pruning: The uncertainty of the group assignment of a structure can be computed as

$$T(s_i) = \frac{\sum_{j=1, j\neq i}^{C} [R'_{ij} \cdot \log(R'_{ij}) + (1 - R'_{ij}) \cdot \log(1 - R'_{ij})]}{C - 1} \tag{3}$$

After computing the uncertainty score [21] for each of the structures, they are sorted based on the descending order of their uncertainty score and then the top 10% are removed from the set as structures with high uncertainty scores are assumed to be of low quality.

3.3 Phase III: Partitioning the Reduced Set of Structures into Groups

In this stage, we combine three group assignment outcomes on the reduced set of structures. At first, phase I is employed to determine the group assignment outcome (with RMSD-based similarity matrix). Then, phase I is replicated with the similarity matrix computed using TM-Score.

For the third group assignment outcome, the work in [4] is utilized which embeds structure-energy pairs in a nearest-neighbor graph (using RMSD to identify nearest neighbors). The reduced set of structures is embedded in a nearest-neighbor graph (nngraph) $G = (V, E)$ where the vertex set V is populated with the structures, and the edge set E is populated by inferring a local neighborhood. The distance between two structures is measured in terms of RMSD after optimal superimposition with respect to a reference structure. The potential energies of the structures are then used to identify groups that form basins in the energy landscape. After building the nngraph, basins are extracted by locating points of attraction or focal minima, identified as vertices with local energy minima in the graph. Each local minimum vertex represents a basin. Once local minima vertices are identified, remaining vertices are assigned to the basins as follows: each vertex u is associated with a negative gradient estimated by choosing the edge (u, v) that maximizes the ratio $[e(u)–e(v)]/d(u, v)$, where $e(u)$ is the energy of the structure in vertex u and $d(u, v)$ denotes the distance between u and v. From each vertex u that is not a local minimum, the negative gradient is followed (via the edge that maximizes the mentioned ratio) until a local minimum is found. Vertices that arrive at the same local minimum are then assigned to the basin/cluster associated with that minimum. The Structural Bioinformatics Library (SBL) [15] is utilized to decompose a structure nngraph into basins/groups.

Now, there are 3 group assignment outcomes on the reduced set of structures (obtained from phase II) and they need to be aggregated into one. As like in phase II, each of the group-label assignments is represented as a vector of the group- labels containing the group-label for each of the structures in the reduced set. Each vector is then converted to a binary similarity matrix M_B in the exact same way the group-assignment resemblance matrix R was formed in phase II. Then, the average similarity matrix $M^{'}$ is computed in the identical manner the average resemblance matrix $R^{'}$ was derived in phase II. Finally, phase I is replicated with this average similarity matrix $M^{'}$ to yield the ultimate group-label assignments of the structures.

3.4 Phase IV: Assigning Score to Each Structure

After figuring out the final group composition, a modified version of the density- based strategy [41] proposed in [2] is adopted to score each structure. Consider that a structure s_i is assigned to a group of l structures, a density score d_i for s_i can be computed,

$$d_i = \frac{\sum_{j=1}^{l} \mathrm{TM} - \mathrm{Score}_{i,j}}{l} \tag{4}$$

where $\mathrm{TM} - \mathrm{Score}_{i,j}$ is the TM-Score between x_i and x_j $(1 \leq i, j \leq l)$. In fact, different metrics can be used but the benefit of considering TM-Score is that its [0, 1] range keeps the density scores in the same range as well.

Groups are sorted (in descending order) by their size (number of structures in the group). Consider that the rank of a group based on size in r_Z. The groups are also sorted (in ascending order) by the average potential energy [8] of the structures in it and let the rank of a group based on that is r_Y. The modified structure density score $d_i^{'}$ is then:

$$d_i^{'} = \frac{d_i}{r_Y + r_Z} \tag{5}$$

The modified scores are leveraged to assign a weight w_i to each structure,

$$\omega_i = e^{d_i^{'}} \tag{6}$$

After assigning weights in this manner, the highest-weight structure is considered as the *best* structure. This score also captures the quality of the structure in the given set of structures.

3.5 Experimental Setup

The proposed method NMF-EL QA is compared to representative state-of-the-art methods: AlphaFold 2 [22, 35] and AlphaFold 3 [1]. The score assigned to the structures is also compared with the actual TM-Score (from the experimentally- known structure in the Protein Data Bank (PDB) [10]) for each protein target.

3.6 Dataset

Table 1. The dataset contains 20 protein targets identified by the corresponding PDB ID (column 2). The chain extracted from a multi-chain PDB entry is shown in parentheses. The sequence length (number of amino acids) of each of the protein targets is on column 3.

#	PDB ID	Sequence Length
1	1AIL	73
2	1WAP(A)	68
3	1TIG	94
4	1DTJ(A)	74
5	1HZ6(A)	64
6	1C8C(A)	64
7	2CI2	83
8	1BQ9	54
9	1HHP	99
10	1FWP	139
11	1SAP	66
12	2H5N(D)	123
13	2EZK	99
14	1AOY	78
15	1ALY	146
16	1CC5	83
17	1ISU(A)	62
18	6MSP	97
19	5UW2	165
20	5JZR	131

The proposed method is evaluated on a dataset containing 20 benchmark proteins of different folds and sequence lengths (number of amino acids). The Rosetta Ab- Initio protocol is utilized to generate 20, 000 all-atom structures for each protein target. It is to be noted that the goal is not just to generate a few structures, as achievable with AlphaFold 2 or AlphaFold 3 now, but to capture a broader view of the structure space by exploring alternative structures.

3.7 Evaluation Metrics

At first, the quality of the scores assigned to structures by the proposed method is evaluated via the Pearson correlation between these scores and the true TM- Score from the experimentally-known structure for each protein target in order to demonstrate the potential of this method for not only single structure selection but also structure quality estimation. Additionally, the top five structures scored by NMF-EL QA are extracted to compare them with the structure outputs yielded by AlphaFold 2 and AlphaFold 3 (both offer five structures for prediction). The average RMSD of those five structures with respect to the experimentally-known structure available in the Protein Data Bank (PDB) [10] is then analyzed with that of the AlphaFold versions.

4 Results

In this segment, two sets of results are presented: the correlation of the assigned scores with the true TM-Score and the comparison of the average RMSD of the top five structures with well-known tools like AlphaFold 2 and 3 on the dataset of Table 1.

4.1 Correlation with Actual TM-Score

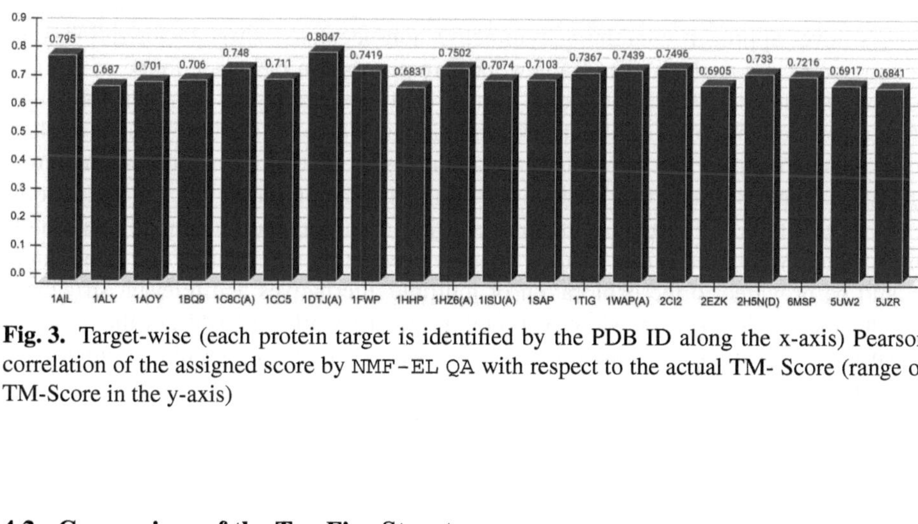

Fig. 3. Target-wise (each protein target is identified by the PDB ID along the x-axis) Pearson correlation of the assigned score by NMF-EL QA with respect to the actual TM- Score (range of TM-Score in the y-axis)

4.2 Comparison of the Top Five Structures

Figure 3 indicates that on 15/20 of the protein targets, NMF-EL QA attains a Pearson correlation coefficient higher than 0.7, representing a strong positive correlation with the true TM-Score. This well-exhibits the potential of this method for not only single structure selection but also structure quality estimation. On the other hand, Fig. 4 demonstrates an additional utility of NMF-EL QA with respect to the state-of-the-art AlphaFold 2 and 3. It indicates that in 6/20 protein targets, the top five structures found by the proposed method are of better quality than both the Alphafold versions, and in 4/20 other cases, the difference in the quality of the structures is negligible.

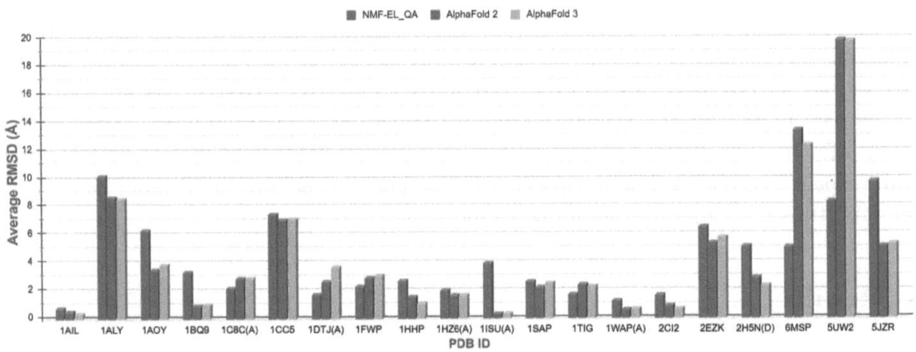

Fig. 4. Comparison of the proposed method NMF-EL QA with Alphafold 2 and Alphafold 3 in terms of the average RMSD (in the y-axis) of the top five structures with respect to the native structure for the corresponding target protein represented by the PBD ID (in x-axis)

5 Conclusion

This paper sets forth a comprehensive structure quality assessment framework that utilizes a novel approach combining symmetric non-negative matrix factorization with energy-landscape analysis. The framework is in possession of a wide range of utilities such as assigning scores to individual structures demonstrating its potential for single structure selection, grouping structures validating its worth for multiple structure selection, aggregating different group assignment outcomes in identifying outlier structures, and so on. The proposed method can also be utilized as a non-parametric clustering tool, making it suitable for structure-function analyses that demand recognizing structural macrostates. While this paper focuses on structure quality assessment, due importance should be provided to the exploration of multiple alternatives and relevant structures beyond the single-structure view of proteins. Future work will address this broader perspective. Furthermore, subsequent studies will explore utilities like summarizing biomolecular dynamics.

References

1. Abramson, J., et al.: Accurate structure prediction of biomolecular interactions with alphafold 3. Nature 1–3 (2024)
2. Akhter, N., Chennupati, G., Kabir, K.L., Djidjev, H., Shehu, A.: Unsupervised and supervised learning over the energy landscape for protein decoy selection. Biomolecules **9**(1), 607 (2019)
3. Akhter, N., Qiao, W., Shehu, A.: An energy landscape treatment of decoy selection in template-free protein structure prediction. Computation **6**(2), 39 (2018)
4. Akhter, N., Shehu, A.: From extraction of local structures of protein energy landscapes to improved decoy selection in template-free protein structure prediction. Molecules **23**(1), 216 (2018)
5. Akhter, N., Vangara, R., Chennupati, G., Alexandrov, B., Djidjev, H., Shehu, A.: Non- negative matrix factorization for selection of near-native protein tertiary structures. In: IEEE International Conference on Bioinformatics and Biomedicine (BIBM), pp. 70–73. San Diego, CA (2019)

6. Akhter, N., et al.: Improved protein decoy selection via non-negative matrix factorization. IEEE/ACM Trans. Comput. Biol. Bioinform. (2021)
7. Alexandrov, L.B., et al.: Signatures of mutational processes in human cancer. Nature **500**(7463), 415–421 (2013)
8. Alford, R.F., Leaver-Fay, A., Jeliazkov, J.R., et al.: The rosetta all-atom energy function for macromolecular modeling and design. J. Chem. Theory Comput. **13**(6), 3031–3048 (2017)
9. Baldassarre, F., Menendez Hurtado, D., Elofsson, A., Azizpour, H.: Graphqa: protein model quality assessment using graph convolutional networks. Bioinformatics **37**(3), 360–366 (2021)
10. Berman, H.M., et al.: The protein data bank. Nucl. Acids Res. **28**(1), 235–242 (2000). https://www.rcsb.org/
11. Boehr, D.D., Wright, P.E.: How do proteins interact? Science **320**(5882), 1429–1430 (2008)
12. Boutsidis, C., Gallopoulos, E.: Svd based initialization: a head start for nonnegative matrix factorization. Pattern Recogn. **41**(4), 1350–1362 (2008)
13. Brooks, B.R., Bruccoleri, R.E., Olafson, B.D., States, D.J., Swaminathan, S.a., Karplus, M.: Charmm: a program for macromolecular energy, minimization, and dynamics calculations. J. Comp. Chem. **4**(2), 187–217 (1983)
14. Brunet, J.P., Tamayo, P., Golub, T.R., Mesirov, J.P.: Metagenes and molecular pattern discovery using matrix factorization. Proc. Nation. Acad. Sci. **101**(12), 4164–4169 (2004)
15. Cazals, F., Dreyfus, T.: The structural bioinformatics library: modeling in biomolecular science and beyond. Bioinformatics **33**(7), 997–1004 (2017)
16. Chen, C., Chen, X., Morehead, A., Wu, T., Cheng, J.: 3D-equivariant graph neural networks for protein model quality assessment. Bioinformatics **39**(1) (2023)
17. Chen, X., Liu, J., Guo, Z., et al.: Protein model accuracy estimation empowered by deep learning and inter-residue distance prediction in CASP14. Sci. Rep. **11** (2021)
18. Cheng, J., Choe, M., Elofsson, A.S., et al.: Estimation of model accuracy in CASP13. Proteins **87**(12), 1361–1377 (2021)
19. Felts, A.K., Gallicchio, E., Wallqvist, A., Levy, R.M.: Distinguishing native conformations of proteins from decoys with an effective free energy estimator based on the opls all-atom force field and the surface generalized born solvent model. Proteins: Struct. Funct. Bioinform. **48**(2), 404–422 (2002)
20. Guo, S.S., Liu, J., Zhou, X.G., Zhang, G.J.: Deepumqa: ultrafast shape recognition based protein model quality assessment using deep learning. Bioinformatics **38**(7), 1895–1903 (2022)
21. Hong, Y., Kwong, S., Chang, Y., Ren, Q.: Unsupervised data pruning for clustering of noisy data. Knowl.-Based Syst. **21**(7), 612–616 (2008)
22. Jumper, J., Evans, R., et al.: Highly accurate protein structure prediction with alphafold. Nature (2021)
23. Kabir, K.L., Akhter, N., Shehu, A.: From molecular energy landscapes to equilibrium dynamics via landscape analysis and markov state models. J. Bioinf. Comput. Biol. **17**(6), 1940014 (2019)
24. Kabir, K.L., Chennupati, G., Vangara, R., Djidjev, H., Alexandrov, B., Shehu, A.: Decoy selection in protein structure determination via symmetric non-negative matrix factorization. In: IEEE International Conference on Bioinformatics and Biomedicine (BIBM), pp. 23–28. Virtual (2020)
25. Kabir, K.L., Akhter, N., Shehu, A.: Connecting molecular energy landscape analysis with markov model-based analysis of equilibrium structural dynamics. In: Proceedings of the 11th Int'l Conference on Bioinformatics and Computational Biology (BICOB), pp. 181–189 (2019)

26. Kabir, K.L., Bhattarai, M., Alexandrov, B.S., Shehu, A.: Single model quality estimation of protein structures via non-negative tensor factorization. In: International Conference on Computational Advances in Bio and Medical Sciences, pp. 3–15 (2021)
27. Kabir, K.L., Hassan, L., Rajabi, Z., Akhter, N., Shehu, A.: Graph-based community detection for decoy selection in template-free protein structure prediction. Molecules **24**(5), 854 (2019)
28. Karasikov, M., Pages, G., Grudinin, S.: Smooth orientation-dependent scoring function for coarse-grained protein quality assessment. Bioinformatics **35**(16), 2801–2808 (2019)
29. Kim, J., Park, H.: Fast nonnegative matrix factorization: an active-set-like method and comparisons. SIAM J. Sci. Comput. **33**(6), 3261–3281 (2011)
30. Kuang, D., Yun, S., Park, H.: Symnmf: nonnegative low-rank approximation of a similarity matrix for graph clustering. J. Glob. Optim. **62**(3), 545–574 (2015)
31. Lazaridis, T., Karplus, M.: Discrimination of the native from misfolded protein models with an energy function including implicit solvation. J. Mol. Biol. **288**(3), 477–487 (1999)
32. Leaver-Fay, A., Tyka, M., Lewis, S.M., Lange, O.F., Thompson, J., Jacak, R., et al.: ROSETTA3: an object-oriented software suite for the simulation and design of macromolecules. Methods Enzymol. **487**, 545–574 (2011)
33. Lei, J., Akhter, N., Qiao, W., Shehu, A.: Reconstruction and decomposition of highdimensional landscapes via unsupervised learning. In: ACM SIGKDD Intl Conf on Knowledge Discovery & Data Mining, pp. 2505–2513. San Diego, CA (2020)
34. Maximova, T., Moffatt, R., Ma, B., Nussinov, R., Shehu, A.: Principles and overview of sampling methods for modeling macromolecular structure and dynamics. PLoS Comp. Biol. **12**(4), e1004619 (2016)
35. Mirdita, M., Schu¨tze, K., Moriwaki, Y., Heo, L., Ovchinnikov, S., Steinegger, M.: Colabfold: making protein folding accessible to all. Nature Methods **19**(6), 679–682 (2022)
36. Mirzaei, S., Sidi, T., Keasar, C., Crivelli, S.: Purely structural protein scoring functions using support vector machine and ensemble learning. IEEE/ACM Trans. Comput. Biol. Bioinf. (2016)
37. Moult, J., Fidelis, K., Kryshtafovych, A., Schwede, T., Tramontano, A.: Critical assessment of methods of protein structure prediction (casp)—round x. Proteins: Struct. Funct. Bioinform. **82**, 1–6 (2014)
38. Olechnoviˇc, K., Monastyrskyy, B., Kryshtafovych, A., et al.: Comparative analysis of methods for evaluation of protein models against native structures. Bioinformatics **35**(6), 937–944 (2019)
39. Pelechrinis, K.: Spectral clustering with eigengap heuristic: A matlab implementation. http://kokkodis.blogspot.com/2013/02/spectral-clustering-with-eigengap.html. Accessed 30 Dec 2024
40. Tunyasuvunakool, K., Adler, J., Wu, Z., et al.: Highly accurate protein structure prediction for the human proteome. Nature **596**, 590–596 (2021)
41. Wang, K., Fain, B., Levitt, M., Samudrala, R.: Improved protein structure selection using decoy-dependent discriminatory functions. BMC Struct. Biol. **4**(1), 1–18 (2004)
42. Zelnik-Manor, L., Perona, P.: Self-tuning spectral clustering. In: Advances in Neural Information Processing Systems, pp. 1601–1608 (2005)
43. Zhang, J., Xu, D.: Fast algorithm for population-based protein structural model analysis. Proteomics **13**(2), 221–229 (2013)

DRPLLM: A Large Language Model-Based Framework for Predicting Drug Response in Cancer Using Multi-Omics Data

Rohan Gnanaolivu[1], Priyanka Vasanthakumari[2], Brenna C. Novotny[1], Jamie C. Overbeek[2], Yuanhang Liu[1], Alexander Partin[2], Nicholas Chia[2], Yitan Zhu[2], and Chen Wang[1](✉)

[1] Department of Quantitative Health Sciences, Mayo Clinic, Rochester, MN 55905, USA
wang.chen@mayo.edu

[2] Data Science and Learning Division, Argonne National Laboratory, Lemont, IL 60439, USA

Abstract. Predicting drug response is critical for advancing precision medicine, as many drugs fail in clinical trials due to a limited understanding of molecular mechanisms. We introduce DRPLLM (Drug Response Prediction using Large Language Model), a novel framework that integrates the Llama- 8b large language model (LLM) with a deep neural network (DNN) regression head. DRPLLM uses embeddings from cell line gene expression, mutations, and drug properties formatted into natural language prompts. Trained on the Genomics of Drug Sensitivity in Cancer v2 and Cancer Cell Line Encyclopedia datasets, DRPLLM achieved a Spearman rank correlation coefficient (SCC) of 0.71, improving to 0.74 after hyperparameter optimization. Validation on a patient- derived xenograft (PDX) cohort showed that the embedding-based approach outperformed the traditional prompt-engineering method SCC = 0.45 ($p < 1e\text{-}16$) versus SCC = 0.06 ($p > 0.05$), demonstrating its potential for accurate, generalizable drug response predictions in precision oncology.

1 Introduction

Cancer is the second leading cause of death worldwide, with conventional treatments such as chemotherapy often yielding suboptimal outcomes and significant side effects [1, 2]. The advent of precision medicine, which incorporates the patient's specific genetic mutations and tailors drug therapies that target molecular pathways involved in disease progression, has shown more promising results with reduced toxicity. In precision oncology, the identification of biomarkers has driven the development of drug response prediction (DRP) models, which help predict how cancer cells will respond to various therapeutic agents [3, 4]. These models facilitate the discovery of more effective and personalized treatments. Publicly available datasets, such as the Cancer Cell Line Encyclopedia [5] (CCLE) and the Genomics of Drug Sensitivity in Cancer [6] (GDSC), have been instrumental in advancing this field by providing comprehensive data on drug-cell line interactions, thereby offering a robust foundation for developing these predictive models.

T. Aldwairi et al. (Eds.): BICOB 2025, CCIS 2535, pp. 189–203, 2025.
https://doi.org/10.1007/978-3-031-94039-2_16

DRP models that are built on conventional neural networks or machine learning models utilize high-dimensional features derived from cancer cell line (CCL) and drug from many CCL-drug pairs with experimentally measured treatment responses. These features include genetic mutations, gene expression profiles, drug SMILES (Simplified Molecular Input Line Entry System) representations, and the physicochemical properties of the drugs [7–10]. By integrating these diverse data types, researchers aim to gain deeper insights into the molecular mechanisms underlying drug responses and to identify novel biomarkers that can guide personalized treatment strategies. However, many conventional DRP models face significant challenges, particularly in terms of overfitting and generalizability [11]. These models often face challenges in handling the complexity of heterogenous features, resulting in reduced generalizability and poor performance on independent datasets. In addition, deep learning approaches of integrating CCL and drug features may encounter challenges of interpretability as the features are generated through non- linear neutral networks and pooling operations.

Large language models (LLMs) are increasingly being leveraged for predictive tasks across various domains, including robotics and precision medicine. These models, such as Gemini [12], Llama-3 [13], and GPT-4 [14], possess billions to trillions of parameters, making them highly capable of understanding and generating human-like language responses. LLMs are trained on vast corpora of text data sourced from books, articles, academic journals, web content, and open-access datasets. This data undergoes extensive preprocessing before being fed into transformer-based architectures, where the models are trained and fine-tuned for specific tasks. LLMs have demonstrated the ability to advance biological and medical interference tasks, with more recent publications showing their capability to predict drug pair synergy [15]. However, the predictive accuracy of LLMs can be influenced by their settings, potentially leading to hallucinations. Extracting embeddings from the final hidden layer offers a way to mitigate this issue, allowing for continuous-scale response prediction.

In this study, we explore Llama-3 8 billion (8b) parameter models potential to predict drug response by leveraging omics features, drug compound characteristics, and combinations of both from CCL-to-drug interaction datasets provided as natural language prompts. Our approach extends DRP beyond in-vitro modeling systems like CCLs, applying it to in-vivo mouse xenografts. These xenografts offer a more accurate representation for therapies such as monoclonal antibodies, where compound-specific information may be limited. Here, we introduce DRPLLM (Drug Response Prediction Large Language Model), a novel model built upon Llama-3 8b, which comprises 8 billion parameters. DRPLLM is integrated with a deep neural network (DNN) regression head that utilizes the embeddings from Llama-3 8b's final hidden layer. The model is trained on 100,000 drug-CCL pairs sourced from the GDSCv2 and CCLE public datasets, using optimized prompts based on omics features. The prompt-based embeddings were selected as the most effective method according to the Spearman rank correlation coefficient (SCC) between predicted and observed drug responses. DRPLLM underwent validation with an independent hold- out test set, demonstrating robustness across various cancer types. By enhancing both prediction accuracy and generalizability, this model shows significant promise as a tool for precision oncology, offering deeper insights into drug response for cancer therapies.

2 Methodology

2.1 Development of DRPLLM

We developed DRPLLM (Fig. 1), a predictive model using data from CCLE and GDSCv2 datasets. To prepare the data, we transformed high-dimensional features into natural language text-based contexts for CCLs and drugs. For CCLs, we used various omics-based features such as gene expression profiles and copy number variant (CNV) status. For drugs, we incorporated gene and biomarker targets, SMILES strings, and molecular properties. We generated three types of prompts: (1) the query with CCL omics context, (2) the query with drug compound context, (3) the query with CCL omics and drug compound context. These prompts were provided as inputs to Llama-3, which has 8 billion parameters, and the embeddings from the final hidden layer were extracted to produce a 3072-dimensional embedding vector for each prompt run on NVIDIA A100 GPU. The embeddings were then evaluated to predict the area under the curve (AUC) continuous measure from the CCLE and GDSCv2 dataset, using multiple regression models, including linear regression, multi-layer perceptron (MLP), XGBoost, and DNN. We selected the DNN model to optimize the hyperparameters (HPO) of the model using the Optuna hyperparameter tuning framework [16] and trained the final model on 80% of the dataset. The remaining 20% was used for validation and testing, with performance evaluated using the SCC metric. There was no overlap between the validation and testing. We applied the same prompt strategy on an independent patient- derived xenograft (PDX) dataset and evaluated the SCC to evaluate the data-dependent features and information gleaned from the embeddings of Llama-3 8b and compared that with a prompt- only response from Llama-3 8b.

2.2 CCL-Drug Dataset and Preprocessing

We sourced data from the CCLE and GDSCv2 as part of the IMPROVE project, which included 10,271 drug-cell line combinations from CCLE and 242,036 from GDSCv2.For the development of DRPLLM, we sampled a total of 100,000 CCL-drug pairs with experimentally measured drug responses. We downloaded gene expression data matrices and copy number variant information for the relevant cell lines. Using an AUC threshold of 0.5, each cell line-drug combination was classified as either resistant or sensitive. We divided the dataset into training, validation, and separate hold-out test sets, splitting the data into 80% training and the remaining 20% as validation and hold-out test sets.

Publicly available gene expression and copy number variation data are considered CCL features. Gene expression data was normalized using Z-scores, and the gene list was filtered to include only those in the Catalogue of Somatic Mutations (COSMIC) [17] database. We selected the top 15 upregulated and downregulated genes to serve as context for the cell line prompts. Copy number data was normalized using a logarithmic transformation, and genes with gains or losses in zygosity were included as context. To represent drug features, we extracted SMILES strings representing the chemical structures of the drugs using RDKit [18]. Description of the drug compounds were obtained from PubChem [19], and used as additional context in the drug prompts.

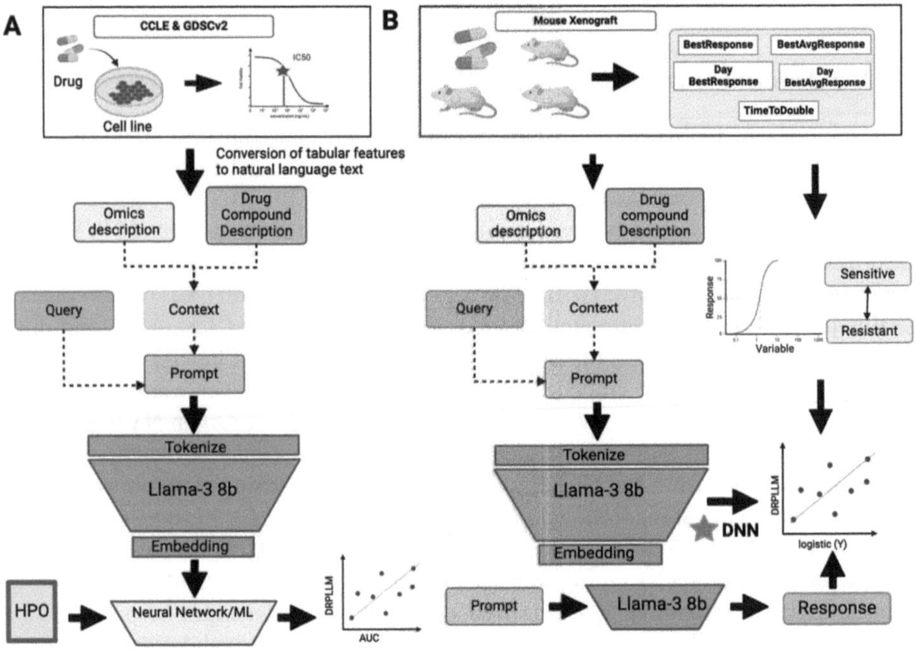

Fig. 1. Method describing the DRPLLM framework. (A) Overview of the drug response prediction process using cell line data from CCLE and GDSC v2. Tabular features such as drug and omics descriptions are converted into natural language text to create a structured query and context. This is tokenized and embedded using the Llama-3 8b model to predict response metrics AUC from an optimized neural network regression head. (B) Workflow for mouse xenograft data involving the transformation of experimental variables, including dose, time to tumor doubling, and sensitivity/resistance metrics, into predicted regression-based coefficients, which serves as the response variable. The drug and omics descriptions are converted into natural language text to create a structured query and context, which is tokenized and embedded using the Llama-3 8b model and the trained deep neural networks (DNN) is used to predict drug response metrics.

2.3 PDX Clinical Trial Dataset and Preprocessing

For the PDX-drug study, a previously published PDX Clinical Trial (PCT) dataset was used [20]. In this dataset, there were a total of 1,075 PDX mouse models where patient tissue was implanted subcutaneously into the flank region of athymic nude female mice using a trocar. The dataset contains 63 drugs and drug combinations, and a total of 4,758 PDX-drug pairs. The PCT dataset categorizes drug-tumor responses in PDX mouse models into four categories: complete responses (CR), partial responses (PR), stable diseases (SD), and progressive diseases (PD). We converted these categories into a continuous scale of probabilities using logistic regression, with response metrics from the data such as "BestResponse", "Day_BestResponse", "BestAvgResponse", "Day_BestAvgResponse", and "TimeToDouble" used as features.

Omics-based features such as gene expression and CNV gene loss or gain using a similar normalizing and selection strategy as the CCL-Drug omics context generation, along with study- specific features, such as tumor size and weight were generated for each PDX xenograft mouse tumor model. Drug compound descriptions were sourced from PubChem; however, for monoclonal antibody drugs absent from PubChem, manual descriptions were obtained via an extensive literature review to capture their role and targets.

2.4 Prompt Creation, Embedding Generation, and Predictors

We developed prompts by transforming omics and drug compound features into natural language text using the DRPLLM preprocessing module. This was achieved through custom Python 3.8 programs designed to convert each data row into meaningful natural language text, leveraging domain-specific knowledge. The data and scripts to generate the prompts are found at https://github.com/dimi-lab/drpllm.

To generate the cell line omics prompts, we first constructed a data table containing cell lines and their associated features. These features included the cell line model ID, tissue site of origin, cancer type, microsatellite instability (MSI) status, The Cancer Genome Atlas (TCGA) cancer type and code, cell culture medium, a list of over and under expressed genes, and genes exhibiting allelic gains or losses. To refine the gene lists specific to cancer, we filtered the expression and copy number variation (CNV) data from GDSCv2 and CCLE to retain only genes found in the Catalogue of Somatic Mutations in Cancer (COSMIC) database. From the expression data, we selected the top 10 upregulated and downregulated genes to be incorporated into the prompt. Similarly, for CNV data, the two most significant genes exhibiting allelic gain or loss were chosen. Once the data table was complete, we formulated natural language sentences that meaningfully represented the values in each row, forming the basis of the cell line prompt.

To generate prompts representing drug compounds, we first retrieved drug information by querying the PubChem database using the drug name or compound ID for all drugs listed in the GDSCv2 and CCLE datasets. This was accomplished using PubChempy v1.0.4. The extracted information included the drug description, molecular weight, and chemical formula. Using RDKit Release_2023_09_4, we generated SMILES strings based on the chemical formula of each drug. A comprehensive data table was then constructed, integrating values obtained from PubChem and RDKit. A structured natural language sentence was formulated for each drug, incorporating these attributes to generate the drug information prompt. The prompts that contained both omics and drug information were concatenated to construct the contextual representation of omics and drug features.

To assess the predictive capabilities of the Llama-3 8b language model embeddings, three types of prompts were developed: (1) query with CCL omics context, (2) query with drug compound context, and (3) query with both CCL omics and drug compound context (Fig. 2). These prompts were processed through Llama-3 8b, generating embeddings from the model's final hidden layer, resulting in a vector size of 3072 for each prompt. The embeddings served as features for training several regression models, including Linear Regression, XGBoost, MLP with default settings, and a custom DNN. The DNN architecture, implemented in PyTorch, consisted of an input layer of size 3072, a flexible series of hidden layers, and an output layer designed to predict a continuous target variable (AUC from the CCLE-GDSCv2 dataset). The model employed Mean Squared

Error (MSE) as the loss function and incorporated early stopping with a patience of 30 epochs to mitigate overfitting. Training was capped at 500 epochs, and performance was evaluated on validation and hold-out test datasets.

In addition, the models were trained on 80% of the dataset, with 20% reserved for validation and testing. For each model, predictions on the test set were evaluated using the SCC as the primary performance metric. Based on model performance, the top-performing model was selected for further hyperparameter optimization, which was found to be the custom DNN. To directly assess the LLMs ability to respond to queries, an additional instruction was appended: "Please provide a single score between 0 and 10, where 0 indicates very likely sensitive and 10 indicates very likely resistant." This prompt format, including query and omics features, was applied to PDX-drug pairs, and the resulting scores were compared to the probabilities derived from the logistic regression model. The SCC metric was then calculated to evaluate the correlation between Llama-3 8B scores and the regression model predictions.

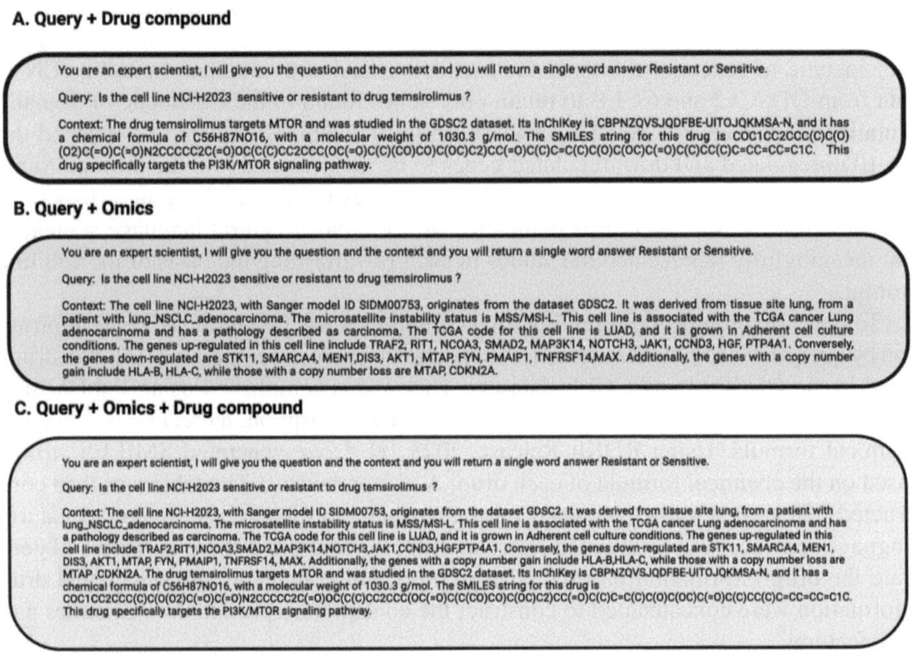

Fig. 2. The figure illustrates three different prompt types utilized in the DRPLLM framework to determine whether the cell line NCI-H2023 is resistant or sensitive to the drug temsirolimus. A. The prompt includes only the query and drug-related context, such as its molecular properties and mechanism of action. B. The prompt integrates the query with detailed cell line omics information, including tissue of origin, microsatellite instability status, gene expression, and copy number variations. C. The most comprehensive prompt, combining the query, omics-based cell line description, and drug-specific details to provide a richer contextual representation for prediction

2.5 Hyperparameter Optimization

To investigate whether optimization of hyperparameters can further enhance the prediction performance of deep-learning models such as DNN, we conducted a hyperparameters search for the DNN using the Optuna framework, optimized only on the validation set to prevent any information leakage into the hold-out test set. Once optimized, the best hyperparameter configuration was applied to the hold-out test set to assess the model's generalization performance. The HPO was conducted over 50 trials, exploring three key hyperparameters: learning rate, batch size, and hidden layer size. For each trial, the DNN was trained for a maximum of 500 epochs, incorporating an early stopping mechanism with a patience of 30 epochs to prevent overfitting. Due to the limited sample size of the PCT cohort, the hyperparameter optimization experiments were focused on the larger CCL-Drug dataset, which combines data from GDSCv2 and CCLE studies. This approach allowed us to rigorously evaluate and optimize the model's performance while ensuring that the hold-out test set was exclusively reserved for final evaluation.

2.6 DRPLLM Comparison with Tabular Based Data Approaches

To evaluate the predictive performance of DRPLLM against traditional machine learning and deep learning models that rely on metric-based features for DRP, we compared its performance to XGBoost, MLP (hidden layers: 128, 64), Random Forest, Linear Regression, and a custom DNN (hidden layers: 512, 128, 64) in predicting AUC. These models were trained using metric-based features derived from 23,565 unique drug–cell line combinations in the GDSCv2 and CCLE datasets. The data was split in an 80:20 ratio and SCC was used as the evaluation metric. To generate the confidence interval bootstrapping by sampling was performed for 1000 iterations. Feature extraction included Morgan fingerprints from SMILES strings, molecular weight of the drug, gene expression levels (transcripts per million) for 1130 cancer know genes, copy number variations for 1126 cancer known genes, and the tissue site of origin, which was label-encoded using Scikit-learn. This process resulted in an initial set of 4311 features, which we refined using feature selection techniques by removing low-variance features (variance < 0.1), which led to a total of 3336 features for model development.

3 Results

3.1 CCL-Drug Response Prediction Results and Hyperparameter Optimization

In this study, we generated embeddings using the Llama-3 8b parameter model for 100,000 drug and cell line pairs using three different prompt types. To evaluate the predictive capability of these embeddings and to estimate which prompt type yielded the highest SCC with experimentally derived AUC metrics from the drug and cell line combination from CCLE and GDSCv2, we employed four machine learning and deep learning models. The performance was assessed using the SCC from the hold-out test set and compared to experientially measured drug response. From our results using the complete data set, we find that the DNN model with the query and omics- only features prompt performed the best, based on the SCC $= 0.71$ (95% confidence interval (CI):

0.701−0.718)(Fig. 3A and Table S1). When stratifying results by cancer type, we found that DNN was the most robust predictive model, outperforming the other three models in eight out of the ten cancer types (Fig.3B and Table S2)

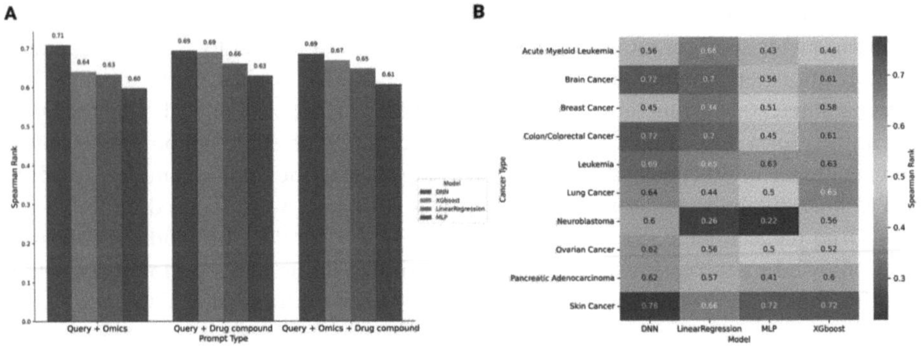

Fig. 3. Performance Comparison of DRPLLM Framework Across Prompt Types, Predictive Models, and Cancer Types. A: Spearman rank correlation coefficient representing the three prompt categories, and their performance with Linear Regression, MLP, XGboost and DNN. B. Heat Map representing the Spearman Rank correlation coefficient from Linear Regression, MLP, XGboost, and DNN from the test data stratified by cancer type

The DNN model was selected as the preferred model of choice due to its superior SCC metrics on the hold-out test set. Optimization of the hyperparameters led to an increase in performance to SCC = 0.74 (95% CI: 0.73–0.75), which is significantly better than the results from the base DNN model, which had an SCC = 0.71 (95% CI: 0.70–0.72) (Table S3). The optimal hyperparameters were learning rate = 0.00013, batch size = 128, and the number of hidden layers = 2, with the number of neurons for each layer being 293 and 853. We find that a lower learning rate yielded higher SCC scores, and the change in batch size did not have an impact compared to the learning rate (Fig. 4).

3.2 Metric Based Model Prediction Performance

To evaluate predictive performance of a metric-based feature generation approach for predicting AUC in drug response modeling, we implemented five regression-based models, which were, Linear Regression, MLP, Random Forest, XGBoost, and DNN. We used the metric SCC to compare their performance. Our analysis revealed that DRPLLM performs comparably to traditional deep learning and machine learning models. However, the DNN model achieved the highest SCC of 0.78 (95% CI: 0.76–0.79), significantly outperforming DRPLLM, which attained an SCC of 0.74 (95% CI: 0.73–0.75) on the test set (Fig. 5A). However, when evaluating the SCC performance stratified by cancer type, DRPLLM demonstrated superior DRP in Brain cancer and Colon/Rectal cancer compared to DNN model trained on tabular metric-based features (Fig. 5B).

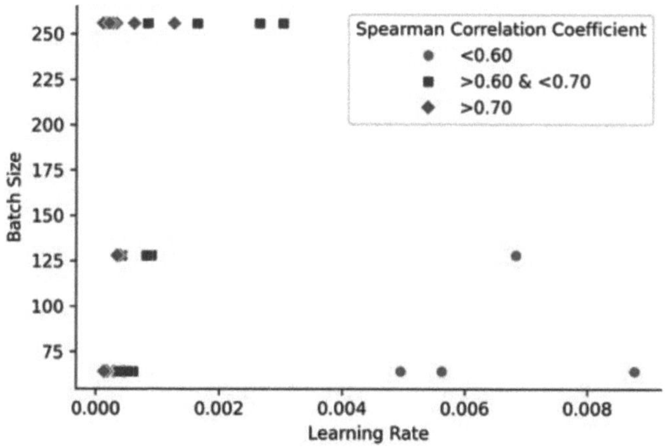

Fig. 4. The relationship between learning rate, batch size, and Spearman Correlation Coefficient in the optimization process of the DNN model using the Optuna hyperparameter framework. Different marker colors represent distinct SCC ranges: red circles for SCC < 0.60, blue squares for SCC between 0.60 and 0.70, and green diamonds for SCC > 0.70.

3.3 PCT Response Data Conversion and Prediction Results

(i) Conversion of PCT response data from categorical to continuous measure: Compared to CCL- drug response data, which were quantified as a continuous score reflective of the area under the dosage and cell viability curve, the PCT response data consisted of four classification categories, making it challenging for constructing a predictive model. To address this issue, we designed a computation strategy to convert PCT response data into a continuous measure. Using the PCT data, we evaluated the predictive power of five response metrics from the dataset: BestResponse, Day_BestResponse, BestAvgResponse, Day_BestAvgResponse, and TimeToDouble. These metrics were used to classify each drug and PDX pair into one of four categories: complete response (CR), partial response (PR), stable disease (SD), or progressive disease (PD). TimeToDouble and Day_BestAvgResponse demonstrated predictive capability across the categories. However, there was significant overlap between them (Fig. 6A). To adapt the dataset for binary classification, aligning it with our training data, we grouped the categories into two classes: *Sensitive* (CR and PR) and *Resistant* (SD and PD). We then evaluated the predictive performance of these five-response metrics for classifying drug-PDX interactions as either resistant or sensitive, based on AUC and AUC-PR (Precision-Recall AUC). Notably, no single metric performed well in predicting this classification (Table 1). However, when we combined all five metrics and applied t-SNE for dimensionality reduction, distinct clustering patterns emerged (Fig. 6B). Finally, a logistic regression model incorporating all five metrics was used to generate predictions on a 0–1 scale. This approach resulted in a close approximation to the original response, achieving an AUC (ROC-AUC) of 0.977 and an AUC-PR of 0.945 (Fig. 6C).

(ii) PDX-drug prediction results: In our study, we aimed to assess the effectiveness of two approaches within our DRPLLM framework using PDX data: (1) a prompting approach, where the LLM was instructed to directly output a score from 1 to 10, indicating

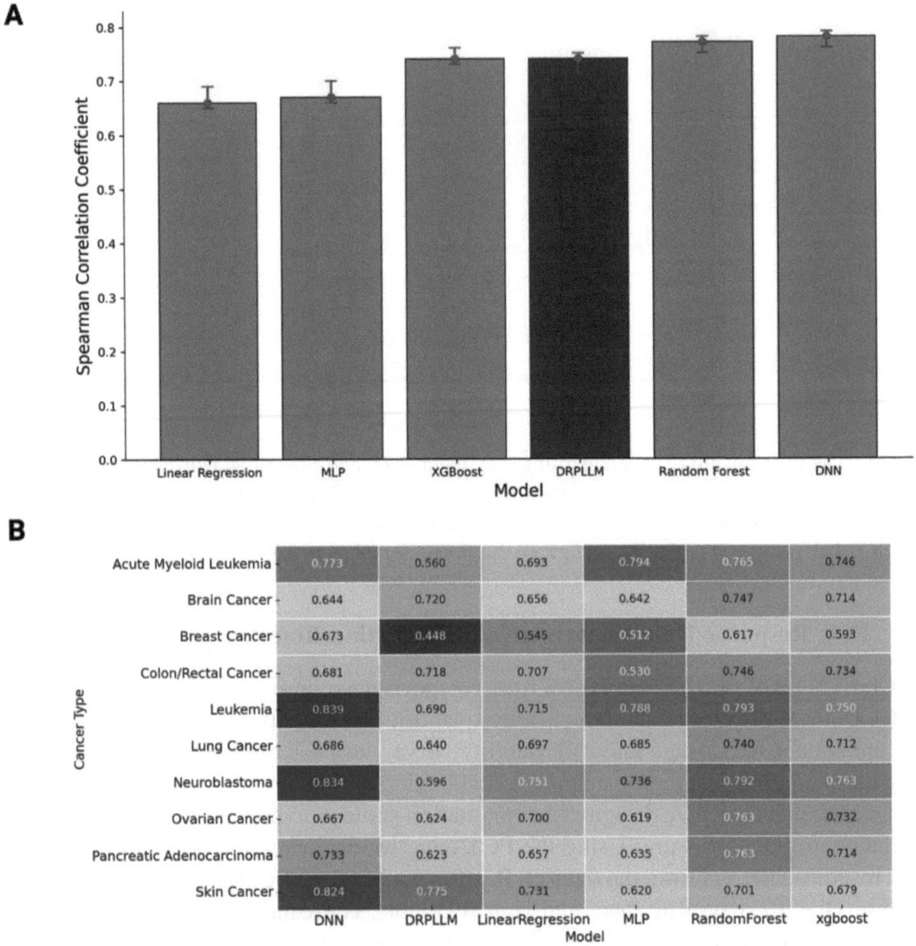

Fig. 5. Performance of structured tabular based features in the development of drug response prediction. A Spearman Correlation Coefficient comparison of DRPLLM (blue) with deep learning and machine learning approaches trained on tabular based features from drug and cell lines listed in the CCLE and GDSCv2 datasets. B. Spearman correlation coefficient results stratified by cancer type for all evaluated deep and machine learning models used in the prediction of drug response from tabular based features.

sensitivity (1 being highly sensitive and 10 being highly resistant), and (2) embedding-based approach, where we extracted embeddings from the Llama-3 8b model's last hidden layer and used them as input features for regression-based predictive models, mirroring our previous work on cell line-drug (CCL-Drug) models.

In the prompting approach, we instructed the LLM to generate direct predictions for drug sensitivity. Unfortunately, this approach performed poorly, yielding a SCC of only 0.0574 (95% CI: –0.1043 to 0.2057), which was statistically insignificant and no better than random predictions. In contrast, the embedding-based approach within

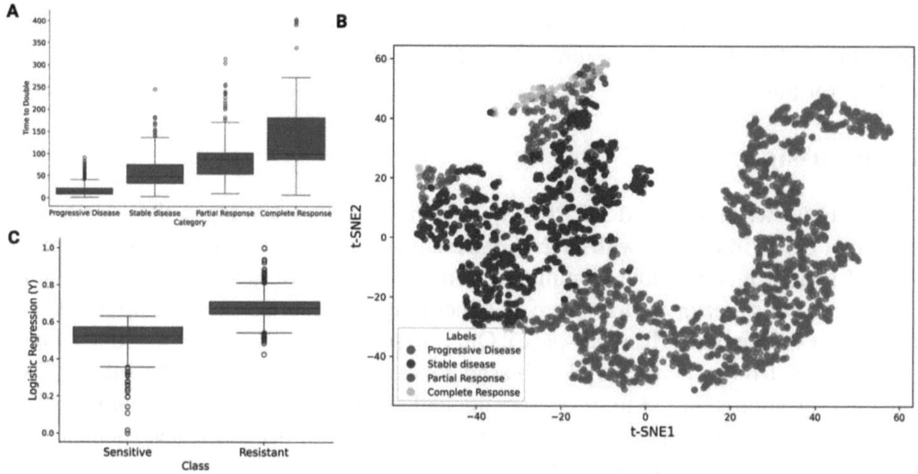

Fig. 6. Conversion of PCT Response Data from Categorical to Continuous Surrogates. A. Box plot of the tumor doubling time versus the four-response category. B: TSNE of PCT data with feature inputs BestResponse, Day_BestResponse, BestAvgResponse, Day_BestAvgResponse, and TimeToDouble. C: Logistic regression of classifying the PCT data from four to two categories Sensitive vs Resistant

Table 1. Area under the curve and area under the precision-recal curve from the response metrics to predict the binary response between the drug and tumor xenograft on the PDX model

Response	*AUC*	*AUC-PR*
BestResponse	0.002354	0.057342
Day_BestResponse	0.775163	0.440364
BestAvgResponse	0.002862	0.054406
Day_BestAvgResponse	0.952186	0.736146
TimeToDouble	0.897865	0.559469

the DRPLLM framework demonstrated substantially better performance. By utilizing embeddings generated from prompts that combined query and omics features, we tested four regression models (DNN, XGBoost, Linear Regression, and MLP) to predict drug response. Among these, the DNN achieved the highest SCC of 0.45 (95% CI: 0.395–0.516), and the results were statistically significant ($p < 1e\text{-}16$) (Table S4). While an SCC of 0.45 may seem modest, it is important to note that in complex biological contexts involving high-dimensional omics data, such values often indicate meaningful predictive ability.

While the straightforward prompting approach presented a convenient solution, it delivered results comparable to random guessing, emphasizing its limitations in capturing meaningful biological relationships. In contrast, the embedding-based strategy, central to the DRPLLM framework, demonstrated statistically significant concordance, with the highest SCC of 0.45. This underscores the robustness of the embedding-based approach, which effectively captures biologically relevant features and offers a scalable solution for integrating diverse data types in oncology research.

4 Discussion

In this study, we introduced DRPLLM, a framework that leverages the Llama-3 8b parameter LLM, integrated with a DNN regression head. After evaluating multiple open-source LLMs, which includes Gemma, Gemini and others, we specifically chose Llama-3 8b due to its superior performance in generating embeddings from multiomics data (results not shown due to limited space). Our comparative analysis revealed that Llama-3 8b produced more diverse and unbiased representations, whereas the other LLMs generated prompts that led to biased results and showed more limited variability. This choice was crucial in our focus on continuous measures of drug efficacy, such as IC50 and drug dosage versus cell viability AUC as drug response measures.

One of the novel aspects of this work was the transformation of tabular data into natural language text, derived from omics and drug features for the CCL/PDX as cancer modeling systems and anti-cancer drugs. The datasets used in this study were key to training and validating the DRPLLM framework across both preclinical and clinical contexts. We evaluated the model on the CCL dataset, which includes 100,000 CCL-drug combinations from GDSCv2 and CCLE studies. We further validated the generalizability of the DRPLLM framework using 3,981 PDX-drug pairs in the PCT cohort. Unlike traditional approaches that rely on structured tabular data, we transformed the multiomic data including gene expression, mutations, and drug properties into natural language prompts and explored the utility of LLM embeddings in predicting drug responses.

Our findings indicate that while deep learning models trained on structured tabular data can outperform embedding generated from prompt-based language models in specific scenarios, LLM embeddings generated from Llama-3 8B achieved comparable and superior performance for certain cancer types. DRPLLM, trained on GDSCv2 and CCLE datasets, initially achieved a SCC of 0.71 (95% CI: 0.701–0.718), which was further optimized to SCC = 0.74 (95% CI: 0.731–0.748). Additionally, when validated on the PDX cohort, the embedding-based approach using Llama-3 8B significantly outperformed traditional zero and few-shot learning prompt-engineering methods, highlighting its potential to improve predictive accuracy in pre-clinical applications.

Importantly, this embedding-based strategy offers a framework for evaluating cohorts where metric-based tabular data is incomplete or unavailable. Traditional deep Learning models use drug representation as morgan fingerprint or drug descriptor metrics encoded in a tabular format [9, 21–23]. However, in the PDX cohort, many drugs were monoclonal antibodies, which lack chemical structures or SMILES representations, making traditional tabular approaches less effective. By leveraging LLM-derived embeddings, DRPLLM overcomes these limitations, offering a generalizable solution for drug response prediction across diverse biological contexts.

This work aligns with the growing use of LLMs in biomedical research, particularly in drug discovery and precision oncology. Previous studies have demonstrated the utility of models like GPT-4 and Gemini for tasks such as drug synergy prediction or binary outcome classification, often using a few-shot learning approach based on prompt creation

[15, 24]. In contrast, our study distinguishes itself by focusing on continuous outcomes of drug efficacy, which provides a more accurate and generalizable representation of experimental data. By converting structured data into natural language, DRPLLM leverages the power of LLM embeddings to capture complex, high- dimensional relationships between cell lines and drugs. This shift toward regression-based prediction, rather than binary classifications, represents a significant advancement in modeling drug responses.

Our study, while promising, is not without limitations. Although the model performed well on the GDSCv2, CCLE, and PDX datasets, its generalizability to broader clinical applications such as patient-derived data from diverse cancer types and treatment modalities requires further validation. Additionally, the embedding-based approach heavily relies on the quality of prompts and features used. Suboptimal prompt design could negatively affect performance, underscoring the need for refining prompt-engineering strategies in future studies. While Llama-3 8b showed clear advantages in this context, the rapidly evolving landscape of LLMs may present newer models that could offer even better performance. Future research should focus on expanding DRPLLM's applications beyond traditional drug response datasets. Integrating more advanced data modalities, such as spatial transcriptomics and single-cell sequencing, could further enhance the model's utility in precision medicine. A direct comparison between LLM-based methods and traditional machine learning approaches will also be important to assess the advantages and trade-offs of using LLMs in terms of computational efficiency, scalability, and interpretability. We also acknowledge that the relatively small dataset imposes constraints on the ability to draw definitive conclusions about the relative strengths of specific prompts or models. Despite these limitations, the statistically significant results achieved by the embedding-based method highlight its reliability and potential for broader applications. Future research should focus on validating DRPLLM with larger datasets, refining prompt engineering to maximize information extraction, and exploring advanced model architectures to further enhance predictive performance.

In summary, DRPLLM demonstrates its effectiveness in predicting drug response by transforming structured data into natural language prompts and leveraging embeddings generated by LLMs. By integrating omics and drug features into a unified framework, DRPLLM offers a novel approach that transcends the limitations of traditional DRP models. The study underscores the potential of LLMs to provide a more nuanced, generalizable, and powerful alternative for handling complex, multiomic data in precision medicine and drug discovery. This study lays a foundation for the integration of LLMs in precision oncology, paving the way for advancements in drug response prediction and personalized treatment strategies. DRPLLM introduces a novel LLM-driven approach to drug response prediction, demonstrating robust performance across CCL datasets. While deep learning models remain competitive, embedding-based methods offer distinct advantages in cases where tabular data is incomplete or unavailable, positioning LLMs as a promising tool for precision oncology applications.

Funding. This research is supported by the Mayo Clinic Center for Individualized Medicine. The funding body played no role in the design of the study and collection, analysis, and interpretation of data and in writing the manuscript.

References

1. Hartl, D., et al.: Translational precision medicine: an industry perspective. J. Transl. Med. **19**(1), 245 (2021). https://doi.org/10.1186/s12967-021-02910-6
2. Collins, D.C., Sundar, R., Lim, J.S.J., Yap, T.A.: Towards precision medicine in the clinic: from biomarker discovery to novel therapeutics. Trends Pharmacol. Sci.Pharmacol. Sci. **38**(1), 25–40 (2017). https://doi.org/10.1016/j.tips.2016.10.012
3. Gadade, D.D., Jha, H., Kumar, C., Khan, F.: Unlocking the power of precision medicine: exploring the role of biomarkers in cancer management. Futur. J. Pharm. Sci. **10** (2024). https://doi.org/10.1186/s43094-023-00573-2
4. Malone, E.R., Oliva, M., Sabatini, P.J.B., Stockley, T.L., Siu, L.L.: Molecular profiling for precision cancer therapies. Genome Med. **12**(1), 8 (2020). https://doi.org/10.1186/s13073-019-0703-1
5. Barretina, J., et al.: The cancer cell line encyclopedia enables predictive modelling of anticancer drug sensitivity. Nature **483**(7391), 603–607 (2012). https://doi.org/10.1038/nature11003
6. Yang, W., et al.: Genomics of drug sensitivity in cancer (GDSC): a resource for therapeutic biomarker discovery in cancer cells. Nucl. Acids Res. **41**(Database issue), D955–61 (2013). https://doi.org/10.1093/nar/gks1111
7. Kuenzi, B.M., et al.: Predicting drug response and synergy using a deep learning model of human cancer cells. Cancer Cell **38**(5), 672–684 e6 (2020). https://doi.org/10.1016/j.ccell.2020.09.014
8. Hostallero, D.E., Li, Y., Emad, A.: Looking at the BiG picture: incorporating bipartite graphs in drug response prediction. Bioinformatics **38**(14), 3609–3620 (2022). https://doi.org/10.1093/bioinformatics/btac383
9. Tang, Y.C., Gottlieb, A.: Explainable drug sensitivity prediction through cancer pathway enrichment. Sci. Rep. **11**(1), 3128 (2021). https://doi.org/10.1038/s41598-021-82612-7
10. Jiang, L., Jiang, C., Yu, X., Fu, R., Jin, S., Liu, X.: DeepTTA: a transformer-based model for predicting cancer drug response. Brief. Bioinform. **23**(3) (2022). https://doi.org/10.1093/bib/bbac100
11. Partin, A., et al.: Deep learning methods for drug response prediction in cancer: predominant and emerging trends. Front. Med. (Lausanne) **10**, 1086097 (2023). https://doi.org/10.3389/fmed.2023.1086097
12. G. Gemini Team, "Gemini: A Family of Highly Capable Multimodal Models," 2024. https://arxiv.org/pdf/2312.11805
13. A.M. Llama Team, "The Llama 3 Herd of Models," 2024. https://arxiv.org/pdf/2407.21783
14. OpenAI, "GPT-4 Technical Report," 2024. https://arxiv.org/pdf/2303.08774
15. Li, T., et al.: CancerGPT for few shot drug pair synergy prediction using large pretrained language models. NPJ Digit. Med. **7**(1), 40 (2024). https://doi.org/10.1038/s41746-024-01024-9
16. Akiba, T., Sano, S., Yanase, T., Ohta, T., Koyama, M.: Optuna: A Next-generation Hyperparameter Optimization Framework, *arXiv,* 2019
17. Tate, J.G., et al.: COSMIC: the catalogue of somatic mutations in cancer. Nucl. Acids Res. **47**(D1), D941–D947 (2019). https://doi.org/10.1093/nar/gky1015
18. "RDKit: Open-source cheminformatics." http://www.rdkit.org
19. Kim, S., et al.: PubChem 2023 update. Nucl. Acids Res. **51**(D1), D1373–D1380 (2023). https://doi.org/10.1093/nar/gkac956
20. Gao, H., et al.: High-throughput screening using patient-derived tumor xenografts to predict clinical trial drug response. Nat. Med. **21**(11), 1318–1325 (2015). https://doi.org/10.1038/nm.3954

21. Cadow, J., Born, J., Manica, M., Oskooei, A., Rodriguez Martinez, M.: PaccMann: a web service for interpretable anticancer compound sensitivity prediction. Nucl. Acids Res. **48**(W1), W502–W508 (2020). https://doi.org/10.1093/nar/gkaa327
22. Greene, C.S., Costello, J.C.: Biologically informed neural networks predict drug responses. Cancer Cell **38**(5), 613–615 (2020). https://doi.org/10.1016/j.ccell.2020.10.014
23. Jin, I., Nam, H.: HiDRA: hierarchical network for drug response prediction with attention. J. Chem. Inf. Model. **61**(8), 3858–3867 (2021). https://doi.org/10.1021/acs.jcim.1c00706
24. Edwards, C., Naik, A., Khot, T., Burke, M., Ji, H., Hope, T.: SynerGPT: in-context learning for personalized drug synergy prediction and drug design, 2023. https://doi.org/10.1101/2023.07.06.547759

Explainable Convolutional Neural Network for COVID-19 Detection

Maxwell Sam[1], Richard Annan[1], Kristen Rhinehardt[2], Kaushik Roy[1], Guoqing Tang[3], and Letu Qingge[1(✉)]

[1] Department of Computer Science, North Carolina A&T State University, Greensboro, NC, USA
{msam,rkannan}@aggies.ncat.edu, {kroy,lqingge}@ncat.edu

[2] Department of Computational Data Science and Engineering, North Carolina A&T State University, Greensboro, NC, USA
klrhineh@ncat.edu

[3] Department of Mathematics and Statistics, North Carolina A&T State University, Greensboro, NC, USA
tang@ncat.edu

Abstract. In the event of the COVID-19 outbreak, prompt and accurate diagnosis became critical for both public health interventions and efficient patient care. COVID-19 is a disease that affects the upper and lower respiratory tract and can have fatal consequences. Early diagnosis is crucial for effective treatment and containment. Studies have shown that COVID-19 manifests in the chest of infected patients, prompting the computer vision community to explore the use of CT scans and deep learning-based solutions for diagnosis. However, efforts to implement explainable artificial intelligence (AI) for interpreting deep learning models in COVID-19 recognition are still scarce. In this paper, we apply SHAP (Shapley Additive Explanations) and LIME (Local Interpretable Model- agnostic Explanations) techniques to enhance the interpretability of our developed CNN that used to detect COVID-19 from CT scan images. The dataset is consist of 4649 images, of which 2476 are from patients with COVID-19 and 2173 are from patients without COVID-19 was used in our implementation. We applied SHAP and LIME techniques to identify important features of COVID-19 images, which even improved the performance of our original model. The comparison results with other baseline models show the robustness of our proposed model and identified important features. We also find that the explainable ability of the SHAP and LIME techniques also depends on its model prediction accuracy.

Keywords: SHapley Additive exPlanations · Local Interpretable Model-agnostic Explanations · COVID-19 Detection · Convolutional Neural Network

This work is supported by the U.S. National Science Foundation under award 2434487 and U.S. National Institutes of Health U24 HG013013.

T. Aldwairi et al. (Eds.): BICOB 2025, CCIS 2535, pp. 204–216, 2025.
https://doi.org/10.1007/978-3-031-94039-2_17

1 Introduction

The COVID-19 pandemic has underscored the critical importance of accurate and timely medical diagnostics. Advanced machine learning techniques, particularly deep learning models, have shown immense potential in aiding the diagnosis of COVID-19 from medical imaging, such as chest X-rays and CT scans [1, 2]. However, the complexity of these models often renders them as "black boxes," where the decision-making process is not transparent to medical practitioners. This lack of interpretability can hinder clinical trust and adoption, as understanding the rationale behind a model's predictions is crucial in the medical domain [3].

To address this challenge, explainable artificial intelligence (XAI) techniques have been developed to provide insights into the decision-making processes of complex models. Among these techniques, SHAP (SHapley Additive exPlanations) and LIME (Local Interpretable Model-agnostic Explanations) have gained prominence [3, 4]. SHAP leverages cooperative game theory to attribute the contribution of each feature towards the model's prediction, ensuring consistency and local accuracy. LIME, on the other hand, approximates the original model locally with an interpretable model, thereby providing insights into individual predictions.

The application of SHAP and LIME to COVID-19 imaging aims to eluci- date how deep learning models derive their conclusions, highlighting the critical features in medical images that drive diagnostic decisions. By doing so, these techniques not only enhance model transparency but also provide a valuable tool for medical professionals to validate and trust machine learning outputs.

The contributions and novelty of this paper are as follows. First, we built and fine-tuned our own developed COVID-CNN model to accurately detect COVID- 19 from CT scan images. Second, we applied XAI techniques, such as SHAP and LIME to identify important features that contributed to the model prediction. Third, we rerun our COVID-CNN model on identified COVID-19 images from XAI techniques and achieve higher prediction accuracy than the original CNN model. Fourth, we compared our CNN model performance with other baseline models, such as VGG-16, VGG-19, Deep-COVID, Deep-COVID DeteCT models. Fifth, we found out although SHAP and LIME identify important features from the input data for the model prediction, the explainable ability of those techniques also depend on model prediction accuracy. If model performance is lower on given dataset, even if we use XAI techniques to identify important features from the input data, that might not be useful and accurate for our prediction task.

The organization of our paper is as follows. In Sect. 2, we review previous related works using model-agnostic techniques to identify relevant features for COVID-19 predictions. Section 3 covers the dataset and provides details on our developed CNN model. In Sect. 4, we apply SHAP and LIME techniques to interpret the model. Section 5 presents the results and Sect. 6 concludes the paper.

2 Related Work

Recent advancements in machine learning (ML) and deep learning (DL) have significantly enhanced computer-aided medical diagnosis, particularly in the analysis of Chest-XR images and CT scans for detecting diseases like COVID-19, pneumonia, and tuberculosis (TB). This overview delves into the latest research findings, focusing on the utilization of medical image analysis, particularly Chest- XR images, to diagnose these critical illnesses.

According to [5], Chest X-ray (CXR) images are considered useful in diagnosing pulmonary disorders such as COVID-19, Pneumonia, and Tuberculosis (TB). A deep learning (DL) model was proposed to enhance disease recognition accuracy while maintaining effective feature extraction. Based on publicly available dataset comprising 7132 CXR images, their model achieved high average test accuracy of 94.31% with a margin of error of about 1.01% whiles a validation accuracy of 94.54% also with a margin of error of about 1.33% through 10-fold cross- validation. Interpretation of their results was facilitated by Gradient-weighted Class Activation Mapping (Grad-CAM), Local Interpretable Model-agnostic Explanation (LIME) and SHapley Additive exPlanation (SHAP). These techniques provided insights into the DL model's decision-making process. Inaddition, they as well made use of eXplainable Artificial Intelligence (XAI) techniques to consolidate and validate model-generated explanations, offering clinicians and medical professionals coherent insights into disease detection and categorization of COVID-19, Pneumonia and Tuberculosis. Authors in [6] also proposed and compared the LIME and SHAP techniques to enhance the interpretation of COVID diagnosis through X-ray scans. In their findings, they first applied SqueezeNet to recognise pneumonia, COVID-19 and normal lung image. Through SqueezeNet, an 84.34% recognition rate success in testing accuracy was obtained. Shapley Additive Explanation (SHAP) and Local Interpretable Model-Agnostic Explanations (LIME) were used to explain and interpret how Squeezenet achieved classification in order to gain a better understanding of what the network observes in relation to a particular task, namely image classification. Their results indicated LIME and SHAP area able to indicate the area of interest and as well help to improve the transparency and the interpretability of the Squeezenet model.

Ashan et al. [7] conducted to detect COVID 19 patients from CXR and CT images and implemented six deep CNN learning models, including VGG16, MobileNetV2, InceptionResNetV2, ResNet50, ResNet101 and VGG19 using 400 CXR and 400 CT images. Achieving an average accuracy of 82.94% on a dataset with CT images and 93.94% on a dataset with CXR images, MobileNetV2 outweigh NasNetMobile. The overall prediction was explained by applying the heatmap of class activation and analyzing the feature extraction implementing LIME. Manjurul et al. [8] achieved high performance on VGG16 model (98.5 $\pm$ 1.19%) among six different Deep CNN models: VGG16, MobileNetV2, InceptionResNetV2, ResNet50, ResNet101 and VGG19 with mixed dataset of CT and X-ray images to classify COVID-19 patients. Results were further explained with LIME. Regarding Sarp et al. [9], they proposed (XAI) technique to detect and interpret COVID-19 positive CXR images. Six deep learning models i.e. VGG16, VGG19, ResNet, InceptionV3, COVID-Net and CORODET were implemented using public dataset collected on GitHub. Their models achieved an accuracy of 96%, 96%, 98%, 91%, 99% and 98% respectively. LIME and SHAP were applied of these models to give a better understanding of the prediction accuracies. To˘ga¸car et al. [10] approach involved three key steps. Firstly, a preprocessing technique to include Fourier Transform and Gradient-weighted Class Activation Mapping to the input images. Secondly, type-based activation sets were generated using three ResNet models before employing the Softmax method. Lastly, the most effective type-based activations are selected using the local interpretable model-agnostic explanations method and re-classified using Softmax. The proposed approach achieved an overall accuracy success of 99.15% across a dataset containing three types of image sets, and a remarkable 99.62% accuracy success specifically for COVID-19 findings. Authors in [11] proposed a comprehensive method for enhancing the interpretability of predictions generated by Convolutional Neural Networks (CNNs) in medical imaging, leveraging Explainable Artificial Intelligence (XAI) techniques. Their approach integrates various techniques, including LIME (Local Interpretable Model Agnostic Explanations), integrated gradients, Anchors and SHAP (Shapley Additive Explanations). Shapley values was used to explain the decisions of their model. Their proposed CNN model achieved a testing accuracy of 90%. XAI techniques was proposed on Deep Learning model for predicting brain tumour status using MRI images data [12]. Their findings presented an interpretable deep learning model for predicting brain tumor types (meningioma, glioma, pituitary) from MRI images. Using a dual-input CNN with Gaussian noise, their model achieved a 94.64% training accuracy and an overall testing accuracy of 85.37%. LIME and SHAP were employed for local and global interpretability to have a clear understanding of their model's prediction. The authors in [13] proposed an explainable framework for detecting spam images using Convolutional Neural Network (CNN) algorithms and Explainable Artificial Intelligence (XAI) algorithms. In their findings their proposed CNN model was used to classify image spam respectively whereas the post-hoc XAI techniques including Local Interpretable Model Agnostic Explanation (LIME) and Shapley Additive Explanations (SHAP) were deployed to provide explanations for the decisions that the black-box CNN models made about spam image detection. Their proposed CNN model achieved a training accuracy of 91% on 6636 image dataset including spam images and normal images collected publicly.

3 Dataset and Our CNN Model

In this study, we fine-tune our own developed convolutional neural network (CNN) model for COVID-19 prediction from CT scan images [14]. Then we utilize SHAP (SHapley Additive exPlanations) and LIME (Local Interpretable Model-agnostic Explanations) for our CNN's model interpretability. SHAP values are computed for each pixel to determine their contribution to the model's output, offering a comprehensive global explanation to the model's predictions. On the other hand, LIME is being leveraged to generate local explanations by perturbing individual images and fitting an interpretable model to approximate the CNN's behavior around each prediction. This results in the identification of superpixels that significantly influence the model's decision.

3.1 The Dataset

The training dataset utilized in this paper was sourced from radiology centers at teaching hospitals in S˜ao Paulo, Brazil [15], and Tehran, Iran [16]. These datasets were combined to form a balanced dataset, designated as "Dataset Mod Dev" for model development. An additional dataset, "Dataset Mod Gen," was compiled from various countries including Russia, China, Italy, Turkey, and Iran [17] to evaluate the model's generalization ability. Dataset Mod Dev contains 4,649 images, with 2,476 from COVID-19 patients and 2,173 from non- COVID-19 patients. Dataset Mod Gen, a more diverse dataset previously used in [17], includes 14,486 images, comprising 7,593 COVID-19 cases and 6,893 non-COVID-19 cases. Figure 1 shows COVID-19 CT scan images as examples, showing patients with and without COVID-19 infection.

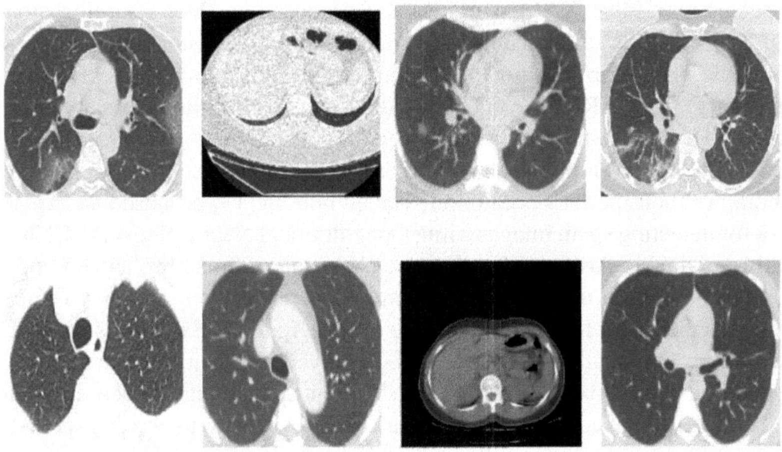

Fig. 1. Example CT-scan of patients with positive COVID-19 in the top row and negative COVID-19 cases in the bottom row.

3.2 The Proposed CNN Model

The COVID-CNN model as shown in Fig. 2 from [14] was used for the application of explainable AI for COVID-19 detection from CT scan images. The COVID-CNN model was specifically designed for grayscale images with dimensions of 300 $\times$ 300 $\times$ 1. The model's processing begins with 116 feature maps (filters) in the first convolutional layer, which has a kernel size of 8 $\times$ 8 and a stride of 22. The output from this initial convolution has dimensions of 97 $\times$ 97 $\times$ 116. Following the convolutional layer, a pooling layer with a stride of 2 $\times$ 2 was applied to downsample the feature maps, reducing the spatial dimensions while preserving key information from the previous layer. These steps were essential for capturing localized and hierarchical patterns in the images. Batch normalization was also applied to the first convolutional layer to stabilize the training process. The second convolutional layer, which also includes batch normalization and max-pooling, mirrors the parameters of the first layer, using 116 filters, an 8×8 kernel, and a 2×2 stride. This results in an output with dimensions of $10 \times 10 \times 116$. This serves as the input to the fully connected layer. The fully connected layer consists of four layers with 362, 184, 78, and 12 neurons, respectively. The ReLU activation function was used in these layers, and dropout regularization applied to prevent overfitting. For the final layer of the COVID- CNN model, the output layer with a softmax activation function and a size of 2 was used for classifying the input image. The model's optimal hyperparameters included a learning rate of 0.001, the ADAM optimizer, and categorical cross-entropy as the loss function.

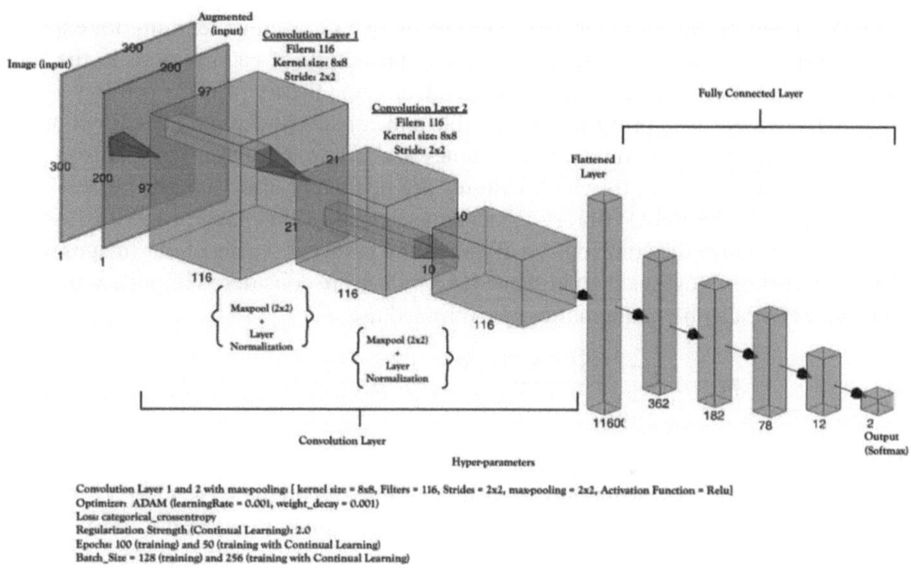

Fig. 2. COVID-CNN model architecture for COVID-19 detection [14].

4 Methodology

Interpretability of Convolutional Neural Network (CNN) on image data has grown to be a critical problem, particularly in high-stakes industries like healthcare, banking, and autonomous systems where decision-making transparency is essential. Notwithstanding their superior performance and accuracy, CNNs are frequently referred to as "black boxes". The term "black box" is mainly because of CNN's opaque and complicated internal workings that makes it challenging to determine which inputs causes which outputs. This lack of interpretability raises challenges in validating, trusting, and deploying CNN models in scenarios where understanding the rationale behind a prediction is as important as the prediction itself. Advanced model interpretability techniques like SHAP (SHapley Additive exPlanations) and LIME (Local Interpretable Model-agnostic Explanations) are used to improve the transparency of CNNs. SHAP leverages game-theoretic principles to assign importance scores to individual regions of an image, offering a global perspective on how each region contributes to the model's overall prediction. This method quantifies the impact of each pixel or superpixel, thereby elucidating the model's decision-making process in a comprehensive manner. Conversely, LIME provides local interpretability by approximating the CNN's behavior with an interpretable surrogate model, specifically tailored to each input image. This approach highlights the critical regions that most significantly influenced the CNN's output, allowing for a more granular understanding of the model's reasoning on a per-instance basis.

4.1 SHAP

SHAP (SHapley Additive exPlanations) is a technique in machine learning for explaining the output of a model by attributing the contribution of each feature to the final prediction. SHAP values are based on cooperative game theory [4]. The primary objective of SHAP values is to provide interpretable and consistent explanations for complex models, thereby enhancing model transparency and trustworthiness. To be able to have much better understanding to which features are positively affecting our models' prediction, we apply the SHAP library to our image data. In this paper, SHAP is used to identify and visualize the contribution of individual pixels or region of our images to the model's prediction. Regions with higher SHAP values are considered important regions. From [4], SHAP can be mathematically expressed as:

$$\phi_i = \sum_{S \subseteq N \setminus \{i\}} \frac{|S|!(|N| - |S| - 1)!}{|N|!} (f(S \cup \{i\}) - f(S)) \quad (1)$$

where: ϕ_i is the SHAP value for feature i, N is the set of all features, S is a subset of features excluding $i, f(S)$ is the model's prediction based on the subset S, $|S|$ is the size of subset S, and $|N|$ is the total number of features that are not included in set S

$$val_x(S) = \int_{\hat{f}(x_1,\ldots,x_p)} dP_{x \notin S} - \mathbb{R}[\hat{f}(X)] \quad [4] \quad (2)$$

4.2 LIME

For image classification tasks, LIME generates local explanations by systematically perturbing the input image and analyzing the model's response to these perturbations to identify which regions of the image are most influential in the model's prediction. LIME for images functions differently compared to its application in tabular or text data. In images, instead of perturbing individual pixels which would have little effect since predictions are typically influenced by larger regions, LIME segments the image into superpixels. A superpixel is a group of connected pixels that share similar properties, such as color or texture. Perturbations are then introduced by randomly masking or turning off these superpixels, often replacing them with a neutral value.

Once the perturbed images are created, the original model is queried to generate predictions for each modified instance. By correlating how the presence or absence of certain superpixels affects the model's output, LIME constructs a linear surrogate model to estimate the contribution of each superpixel to the final prediction. This locally interpretable model assigns importance scores to superpixels, allowing for a clear visualization of the most critical regions that influenced the model's decision. The use of superpixels ensures that explanations are meaningful, as they capture the collective contribution of spatially coherent regions rather than individual pixels, which might not independently influence the outcome. The entire process of LIME on images can be expressed as:

$$\arg\min_{g \in G} \sum_{i=1}^{m} \pi_x(z_j')(f(h_x(z_j')) - g(z_j'))^2 + \Omega(g) \quad (3)$$

where:

- $f(x_i')$: Prediction of the black-box model for the perturbed image x_i'.
- $g(z_j'; \mathbf{w})$: Prediction of the surrogate model (linear model) for the binary vector representation z_j' of the perturbed image with weights $\mathbf{w}$.
- $\pi_x(x_i')$: Kernel function (Proximity score) that gives higher weights to perturbed images close to the original image x.
- $\Omega(g)$: Regularization term to enforce sparsity in the surrogate model for interpretability.

4.3 Intersection Between LIME and SHAP

Both LIME and SHAP techniques aim is to provide insights into the model's decision-making process, but they do so from different perspectives and through different mechanisms. In image analysis, the intersection of LIME and SHAP entails integrating their advantages to produce a more reliable interpretation to a model's unique prediction. Consistent patterns of feature importance can be found by comparing the explanations offered by LIME and SHAP. This increases confidence in the interpretation of the model. For example, when the same superpixels or regions in an image are frequently highlighted by both LIME and SHAP as being crucial for a prediction, it strengthens the confidence in the significance of those regions.

5 Results and Discussion

As shown in Fig. 3, the application of the SHAP library to the COVID-19 CT scan data reveals the specific regions and features that exert the greatest influence on the model's prediction. SHAP quantifies the contribution of each region of our image data to the model's prediction by assigning Shapley values. These values indicate how much each region adds to or detracts from the classification decision. In essence, SHAP breaks down the CNN's prediction into parts, showing which regions have the most significant positive or negative impact on the final output. This detailed breakdown allows for a "region-level" understanding of the model's inner workings. By highlighting the most influential areas, SHAP helps us see exactly what the CNN is focusing on. The highlighted regions in

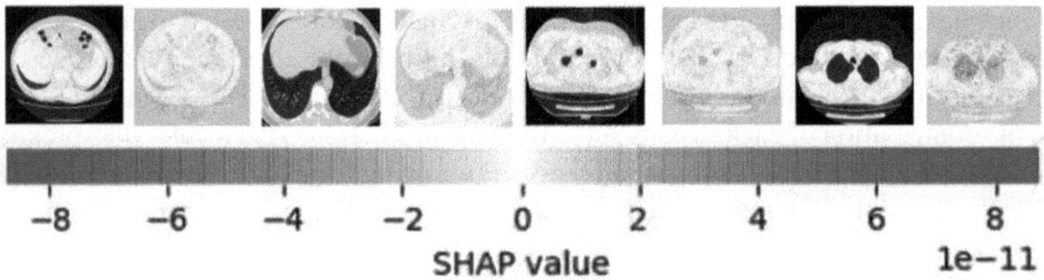

Fig. 3. CT scan images with SHAP explanations for COVID-19 classification, highlighting blue regions as less important and red regions as important for increasing the likelihood of a positive prediction.

red represent the areas that have the most significant influence on the CNN's predictions, indicating that these regions are critical for the model's decision-making process. In contrast, the blue-highlighted areas correspond to regions that have little to no impact on the model's prediction, suggesting that they are not relevant for the classification outcome.

Figure 5 demonstrates the application of LIME (Local Interpretable Modelagnostic Explanations) on our COVID-19 CT scan images. The yellow contours overlaid on the CT images highlight the regions that are of critical importance to the model's prediction. Specifically, LIME enables identify localized areas within the images that contributed the most to the model's decision.

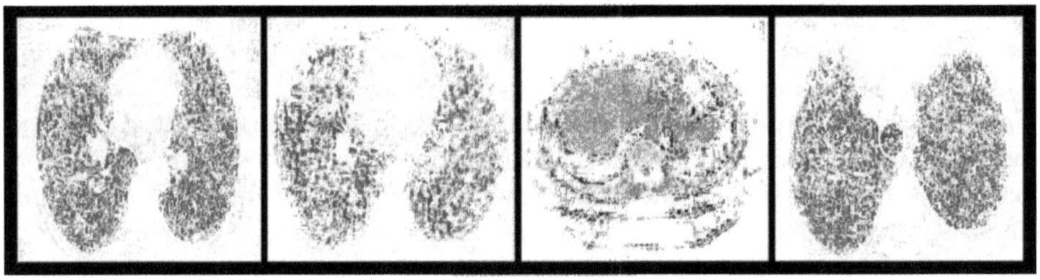

Fig. 4. CT scan slices after applying SHAP, highlighting the regions most critical to the model's prediction.

After thoroughly identifying the most important features using SHAP along with LIME, we carefully modified the original CT scan images by masking out the remarkably non important regions. The areas carefully masked in black, per Fig. 4, clearly show the important regions SHAP identified. Afterward, the CNN model was retrained with the generated images that only had the most important features. As presented in Table 1, we applied SHAP and LIME, to the entire dataset of both positive and negative COVID-19 images. The generated images were then divided into training and testing datasets. After training our model on these processed images, the SHAP identified important regions yielded an accuracy of 96.13%, while the LIME identified regions achieved an accuracy of 95.47%. The intersection of both SHAP and LIME-identified regions recorded an accuracy of 89.78%. These results underscore the model's reliance on the key regions

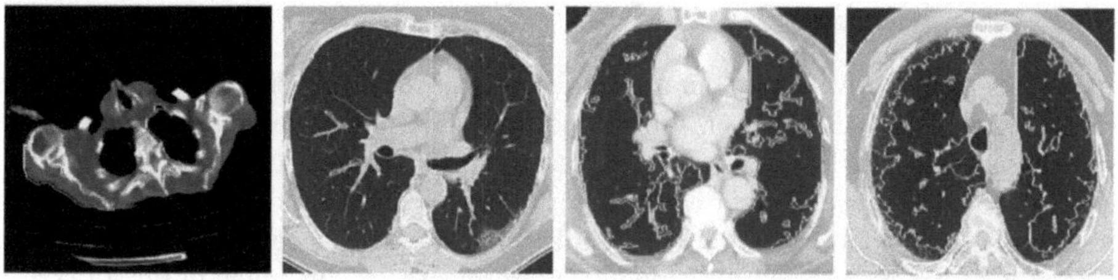

Fig. 5. CT scan images displaying LIME-generated explanations for COVID-19 classification, where yellow contours highlight key regions identified as significant contributors to the model's prediction.

for decision-making, as removing non-influential features did not significantly impact its performance. This further validates the effectiveness of XAI-based feature selection in enhancing model interpretability without compromising predictive accuracy.

5.1 Comparison Results

We compared our results with other models for the detection of COVID-19 from CT scan images such as Deep COVID DeteCT [18], VGG16 [19], VGG19 [19] and and Deep-COVID [16]. These Deep COVID DeteCT [18] and Deep-COVID [16] models utilized famous InceptionV3 [18] and NASNetLarge [16] pre-trained weights techniques and demonstrated substantial testing accuracy when trained on the dataset, mod dev dataset. However, applying these interpretability techniques, SHAP and LIME showed a decline in their accuracies. From Table 1 and Fig. 6, we can see that our COVID-CNN model [14] used for the prediction of COVID-19 CT scans performed better compared with other existing models when SHAP, LIME, and the intersection between both techniques were applied. This demonstrates the effectiveness of interpretability in highlighting important features, thereby optimizing model accuracy and reliability.

Table 1. Comparison results with existing models

Models	SHAP		LIME		Intersection		All Regions	
	Accuracy %	F1-Score	Accuracy %	F1-Score	Accuracy %	F1-Score	Accuracy %	F1-Score
COVID-CNN (Ours)	96.13	96.13	95.47	95.47	89.18	88.72	97.85	97.85
VGG19 [19]	90.37	90.31	93.83	93.82	84.24	84.21	88.55	87.42
VGG16 [19]	92.26	91.80	89.94	89.40	76.43	76.10	87.12	86.43
Deep COVID DeteCT [18]	86.23	69.99	69.42	71.49	70.39	70.21	70.89	70.08
Deep-COVID [16]	74.62	74.40	78.25	77.60	72.31	71.00	95.59	84.67

5.2 Evaluation Metrics

To evaluate the effectiveness of our model, Accuracy, Recall, Precision and the F1-score are used [20]. Accuracy represents the proportion of correct predictions out of the total predictions made. The F1-score, calculated as the harmonic mean of precision and recall, ensures a balance between precision which is the correctness of positive predictions and recall applies to the model's ability to identify all positive instances. These metrics offer a comprehensive evaluation of the model's performance, as illustrated in Eq. 4 [21].

$$
\begin{aligned}
\text{Accuracy} &= \frac{TP + TN}{TP + TN + FP + FN} \\
\text{Precision} &= \frac{TP}{TP + FP} \\
\text{Recall} &= \frac{TP}{TP + FN} \\
\text{F1 - score} &= \frac{2 \times \text{Precision} \times \text{Recall}}{\text{Precision} + \text{Recall}}
\end{aligned}
\tag{4}
$$

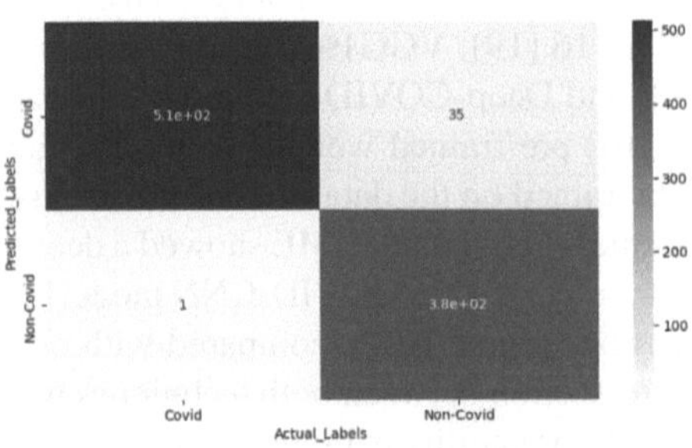

(a) Confusion Matrix for SHAP Generated 4649 Images

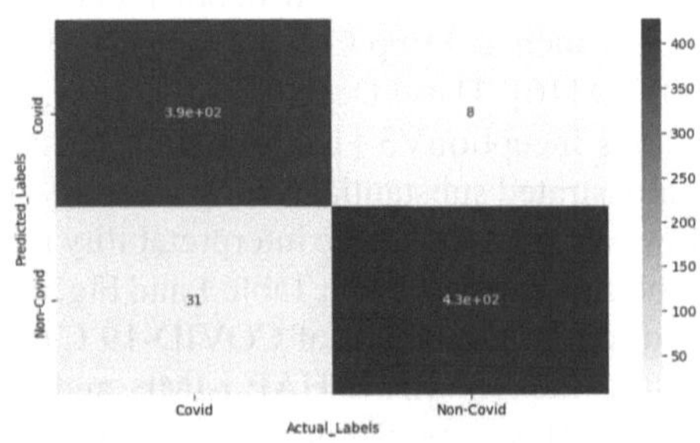

(b) Confusion Matrix for LIME Generated 4649 Images

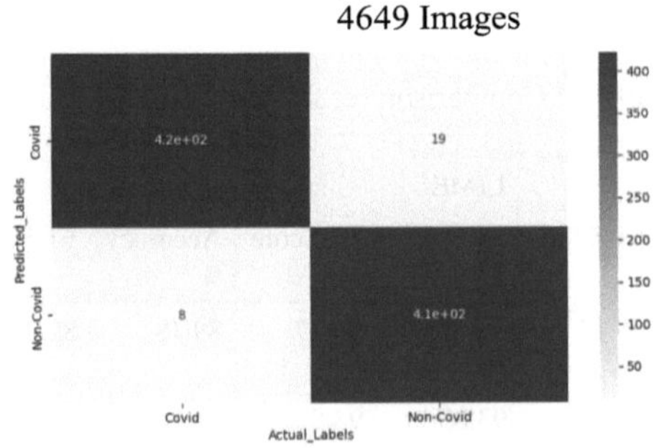

(c) Confusion Matrix for Intersection between SHAP and LIME on 4649 Generated Images

Fig. 6. Confusion Matrix for SHAP, LIME, and Intersection Using our CNN Model

6 Conclusion

In this paper, we applied explainable AI techniques SHAP and LIME to our developed COVID-CNN model [14] specifically developed to predict COVID-19 from CT scan images. After applying SHAP and LIME to our adopted model, the results provide valuable insights into the model's decision-making process. SHAP captured feature importance, demonstrating how the model focused heavily on specific regions where COVID-19 abnormalities typically occur. LIME, on the other hand, offered local interpretability by highlighting features critical to the predictions. The intersection features between SHAP and LIME confirmed consistent feature attributions, demonstrating features that are consistent to both interpretability techniques. Using all images of important regions identified by both SHAP and LIME, we retrained our adopted model using only the important regions masking out the non-important regions. Retraining on SHAP identified regions resulted in a 96.13% accuracy, while LIME based regions yielded 95.47% accuracy. This demonstrated how focusing only on relevant features reduces overfitting and model complexity. Intersection between both SHAP and LIME achieved an accuracy of 89.18%. The model retrained on SHAP identified regions achieved a prediction accuracy of 49.14% when tested on the original images, while the model retrained on LIME based regions attained an accuracy of 50.54%. This depends on setting a threshold $\in$ to determine which SHAP values are considered significant. We set $\in$ equal to or greater than 1e-5 for both SHAP and LIME to filter out insignificant feature contributions.

Although SHAP and LIME identify important features from the input data for the model prediction, the explainable ability of those techniques also depend on model prediction accuracy. If model performance is lower on given dataset, even if we use XAI techniques to identify important features from the input data, that might not be useful and accurate. All implementations developed in this paper can be found at: https://github.com/kobinasam/Explainable-Convolutional-Neural-Network-for-COVID-19-Images (Fig. 6).

Acknowledgment. This work is supported by the U.S. National Science Foundation under award 2434487 and U.S. National Institutes of Health U24 HG013013. We thank anonymous reviewers for their insightful comments and inputs.

References

1. Ng, M.-Y., et al.: Imaging profile of the covid-19 infection: radiologic findings and literature review. Radiol. Cardiothorac. Imaging **2**(1), e200034 (2020)
2. Wang, L., Lin, Z.Q., Wong, A.: Covid-net: a tailored deep convolutional neural network design for detection of covid-19 cases from chest x-ray images. Sci. Rep. **10**(1), 19549 (2020)
3. Ribeiro, M.T., Singh, S., Guestrin, C.: Why should i trust you?: explaining the predictions of any classifier. In: Proceedings of the 22nd ACM SIGKDD International Conference on Knowledge Discovery and Data Mining, ACM, pp. 1135–1144, 2016
4. Lundberg, S.M., Lee, S.-I.: A unified approach to interpreting model predictions. In: Advances in Neural Information Processing Systems, NIPS, pp. 4765–4774, 2017
5. Bhandari, M., Shahi, T.B., Siku, B., Neupane, A.: Explanatory classification of cxr images into covid-19, pneumonia and tuberculosis using deep learning and xai. Comput. Biol. Med. **150**, 106156 (2022)

6. Ong, J.H., Goh, K.M., Lim, L.L.: Comparative analysis of explainable artificial intelligence for covid-19 diagnosis on cxr image. In: 2021 IEEE International Conference on Signal and Image Processing Applications (ICSIPA), pp. 185–190, 2021
7. Ahsan, M.M., Gupta, K.D., Islam, M.M., Sen, S., Rahman, M.L., Shakhawat Hossain, M.: Covid-19 symptoms detection based on nasnetmobile with explainable ai using various imaging modalities. Mach. Learn. Knowl. Extr. **2**(4), 490–504 (2020)
8. Ahsan, M.M., Nazim, R., Siddique, Z., Huebner, P.: Detection of covid-19 patients from ct scan and chest x-ray data using modified mobilenetv2 and lime. Healthcare **9**(9) (2021)
9. Sarp, S., et al.: An xai approach for covid-19 detection using transfer learning with x-ray images, Heliyon (2023)
10. Togacar, M., Muzoglu, N., Ergen, B., Yarman, B.S.B., Halefoglu, A.M.: Detection of covid-19 findings by the local interpretable model-agnostic explanations method of types-based activations extracted from cnns. Biomed. Signal Process. Control **71**, 103128 (2022)
11. Abeyagunasekera, S.H.P., Perera, Y., Chamara, K., Kaushalya, U., Sumathipala, P., Senaweera, O.: Lisa: enhance the explainability of medical images unifying current xai techniques. In: 2022 IEEE 7th International conference for Convergence in Technology (I2CT), pp. 1–9, 2022
12. Gaur, L., Bhandari, M., Razdan, T., Mallik, S., Zhao, Z.: Explanation-driven deep learning model for prediction of brain tumour status using mri image data. Front. Genet. **13**, 822666 (2022)
13. Zhang, Z., Damiani, E., Hamadi, H.A., Yeun, C.Y., Taher, F.: Explainable artificial intelligence to detect image spam using convolutional neural network. In: 2022 International Conference on Cyber Resilience (ICCR), pp. 1–5, 2022
14. Annan, R., Qin, H., Qingge, L.: Generalized deep learning models for COVID-19 detection with transfer and continual learning. In: Proceedings of the 16th International Conference on, vol. 101, pp. 58–72, 2024
15. Soares, E., Angelov, P., Biaso, S., Froes, M., Abe, D.: Sars-cov-2 ct-scan dataset: a large dataset of real patients CT scans for sars-cov-2 identification, 2020
16. Ghaderzadeh, M., Asadi, F., Jafari, R., Bashash, D., Abolghasemi, H., Aria, M.: Deep convolutional neural network–based computer-aided detection system for COVID-19 using multiple lung scans: design and implementation study, vol. 23, p. e27468, 2021
17. Maftouni, M., Law, A.C.C., Shen, B., Kong, Z.J., Zhou, Y., Yazdi, N.A.: A robust ensemble-deep learning model for covid-19 diagnosis based on an integrated ct scan images database. In: IIE Annual Conference. Proceedings, Institute of Industrial and Systems Engineers (IISE), pp. 632–637, 2021
18. Lee, E., et al.: Deep covid detect: an international experience on covid-19 lung detection and prognosis using chest CT. NPJ Digit. Med. **4**, 11 (2021)
19. Karim, M.R., Dohmen, T., Cochez, M., Beyan, O., Rebholz-Schuhmann, D., Decker, S.: Deepcovidexplainer: explainable COVID-19 diagnosis from chest x-ray images. In: 2020 IEEE International Conference on Bioinformatics and Biomedicine (BIBM), pp. 1034–1037, 2020
20. Yang, P., Sturtz, J., Qingge, L.: Progress in blind image quality assessment: a brief review. Mathematics **11**(12) (2023)
21. Miao, J., Zhu, W.: Precision–recall curve (PRC) classification trees. Evol. Intel. **15**(3), 1545–1569 (2022)

Author Index

T. Aldwairi et al. (Eds.): BICOB 2025, CCIS 2535, pp. 217–218, 2025.
https://doi.org/10.1007/978-3-031-94039-2

MIX
Papier aus verantwortungsvollen Quellen
Paper from responsible sources
FSC® C105338

If you have any concerns about our products,
you can contact us on
ProductSafety@springernature.com

In case Publisher is established outside the EU,
the EU authorized representative is:
Springer Nature Customer Service Center GmbH
Europaplatz 3, 69115 Heidelberg, Germany

Printed by Libri Plureos GmbH
in Hamburg, Germany